High-Efficiency Gas Furnace Troubleshooting Handbook

BILLY C. LANGLEY

PRENTICE HALL, Englewood Cliffs, New Jersey 07632

Library of Congress Cataloging-in-Publication Data

LANGLEY, BILLY C.
　　High efficiency gas furnace troubleshooting handbook / Billy C.
Langley.
　　　　p. cm.
　　Includes index.
　　ISBN 0-13-388653-0
　　1. Furnaces—Maintenance and repair. 2. Air conditioning—
Equipment and supplies—Maintenance and repair. 3. Refrigeration
and refrigerating machinery—Maintenance and repair. 4. Gas
appliances. I. Title.
TH7623.L36 1991
697′.07—dc20
　　　　　　　　　　　　　　　　　　　　　　　　　　　　90-49387
　　　　　　　　　　　　　　　　　　　　　　　　　　　　CIP

Acquisitions editor: Susan Willig
Editorial/production supervision: Merrill Peterson
Interior design: Joan Stone
Cover design: Lundgren Graphics Ltd.
Prepress buyer: Mary McCartney
Manufacturing buyer: Ed O'Dougherty

© 1991 by Prentice-Hall, Inc.
A Division of Simon & Schuster
Englewood Cliffs, New Jersey 07632

Printed in the United States of America
10 9 8 7 6 5 4 3 2 1

ISBN 0-13-388653-0

Prentice-Hall International (UK) Limited, *London*
Prentice-Hall of Australia Pty. Limited, *Sydney*
Prentice-Hall Canada Inc., *Toronto*
Prentice-Hall Hispanoamericana, S.A., *Mexico*
Prentice-Hall of India Private Limited, *New Delhi*
Prentice-Hall of Japan, Inc., *Tokyo*
Simon & Schuster Asia Pte. Ltd., *Singapore*
Editora Prentice-Hall do Brazil, Ltda, *Rio de Janeiro*

Contents

6

The Coleman T.H.E. Model B High-Efficiency Gas Furnace   193

7

Electronic Ignition Systems   226

Preface

High-Efficiency Gas Furnace Troubleshooting Handbook was written for the more advanced student in the heating, ventilating, air conditioning, and refrigeration industry and for experienced technicians who want to increase their knowledge in this field. The newest available information was used so that those wanting to learn about this type of equipment would have an opportunity to learn about the newest equipment. Even though some of the control systems used by some manufacturers are different, the basic theory remains the same and is invaluable to those wanting to learn the latest information, provide better service to their customers, and gain advancement in this industry.

Chapter 1 is an overview of the required venting practices and codes for high-efficiency gas furnaces. The types of heat exchangers and their purpose are covered. General installation requirements of the high-efficiency gas furnace electrical and gas piping are discussed to aid the technician in completing these operations as economically as possible.

Chapter 2 presents the Carrier (58) high-efficiency gas furnace. Specific service, installation, and maintenance requirements are presented for this brand of furnace.

Chapter 3 is a presentation of the Heil gas furnace. Specific service, installation, and maintenance requirements pertaining to this furnace are presented so that peak efficiency of the equipment can be maintained.

Chapter 4 covers the Lennox Pulse™ GSR14 gas furnace. Specific service, installation, and maintenance procedures are presented to aid the technician in understanding this type of furnace so that economical operation can be obtained.

Chapter 5 is a presentation of the Coleman T.H.E. gas furnace, models ---- (Dash) and A models only. Their operation, maintenance, service, and installation are covered to aid the technician in ecnomically and successfully completing these procedures.

Chapter 6 is a presentation of the Coleman T.H.E. model B high-efficiency gas furnace. The installation, maintenance, and service procedures for this model furnace are covered in detail to aid the technician in accomplishing these procedures.

Chapter 7 is a presentation of the theory of electronic ignition systems used on the high-efficiency gas furnaces. Flame rectification principles are covered in everyday language to make them easier to understand. The procedures used to check out Honeywell intermittent ignition systems (IID) are presented. Troubleshooting Honeywell intermittent pilot systems are presented to aid the technician in learning these procedures. Also covered are checkout and troubleshooting procedures for the Honeywell direct ignition systems used on high-efficiency gas furnaces.

Where applicable, all chapters include troubleshooting charts and procedures for the units being presented. This book should be of great use to all personnel in the heating industry. My best wishes go with you in all of your endeavors.

Billy C. Langley

1

High-Efficiency Gas Furnaces

The need to conserve energy has caused most builders to use more and better insulation, to include better vapor barriers around the buildings, to use weather stripping, to use better construction techniques to improve the tightness of buildings, and to take other energy conservation measures. The buildings, as a result, are more energy efficient, due in part to less air infiltration into the building. This reduction in air infiltration has resulted in the furnaces and other gas appliances being starved for combustion air, causing a need for a more positive combustion air supply. The high-efficiency gas furnaces provide this necessary air with a combustion air fan or inducer.

HIGH-EFFICIENCY GAS FURNACES

The desire to conserve energy has also resulted in more efficient gas furnace designs. Gas furnaces that are designed to produce 90% and upwards efficiency are known as *condensing gas furnaces*. The higher efficiency is obtained by condensing the water vapor from the flue gases and transferring the latent heat given up to the circulating air for the inside of the building. The water vapor is condensed in a device called a secondary heat exchanger.

Secondary Heat Exchangers

There are several methods used to provide a secondary heat exchanger, such as a finned coil and additional heat exchangers.

Finned coils. The finned coil is located in the air stream ahead of the regular heat exchanger (see Figure 1.1). The products of combustion first enter the primary heat exchanger from the combustion zone and heat is transferred to the circulating air the same as in a standard gas heating furnace. The products of combustion then pass through a tail pipe, or other type of design, into

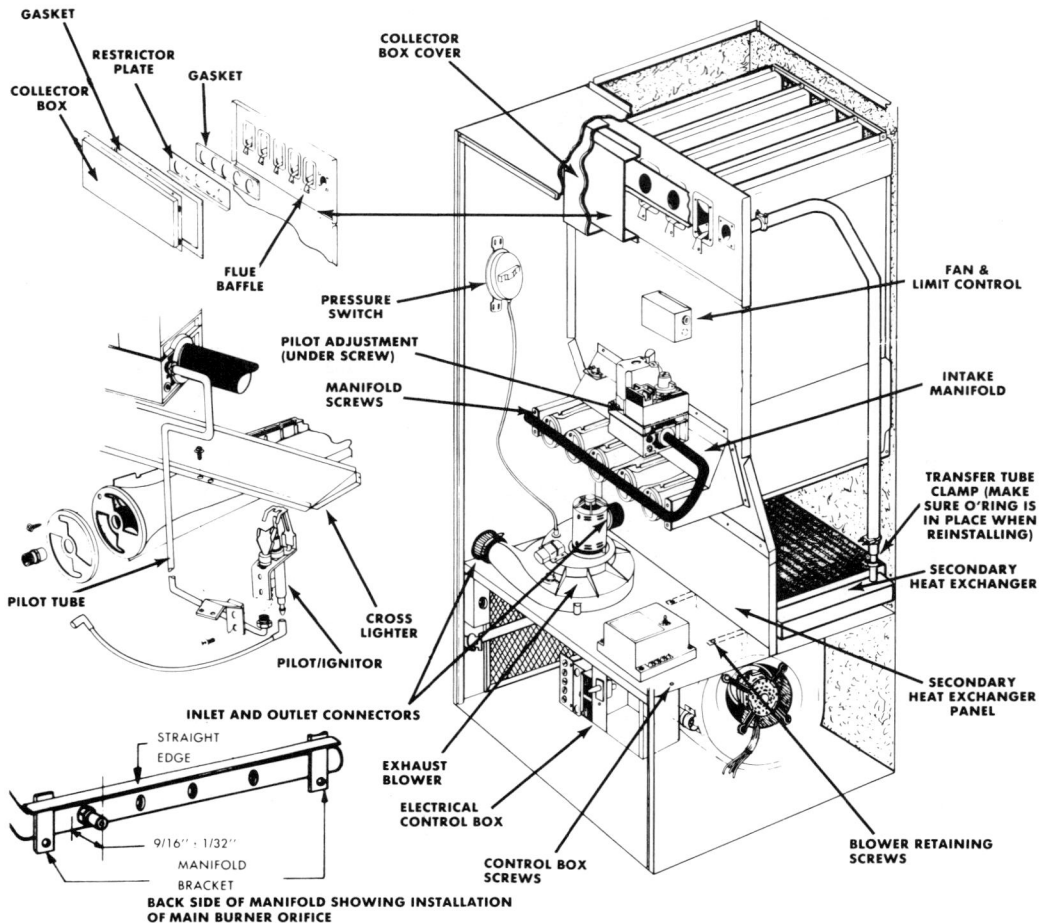

Figure 1.1 Finned coil secondary heat exchanger location. (Courtesy of Heil-Quaker Corporation.)

the secondary heat exchanger. The cooler air passes over this heat exchanger first and causes the water vapor to condense, giving up the latent heat to the circulating air. The circulating air then flows over the warmer primary heat exchanger where it absorbs more heat.

When an additional heat exchanger is used the products of combustion pass from the primary heat exchanger to the secondary heat exchanger. Some of the circulating air passes over the primary heat exchanger and some passes over the secondary heat exchanger (see Figure 1.2). The circulating air does not pass through both the primary and secondary heat exchangers on each pass through the furnace in this type of furnace. Some manufacturers use two secondary heat exchangers. The first one increases the efficiency to the 80 to 85% efficient range and the second one increases the efficiency to the 95 to 98% efficient range. When only one secondary heat exchanger of this type is used the furnace is not generally considered to be a condensing gas furnace.

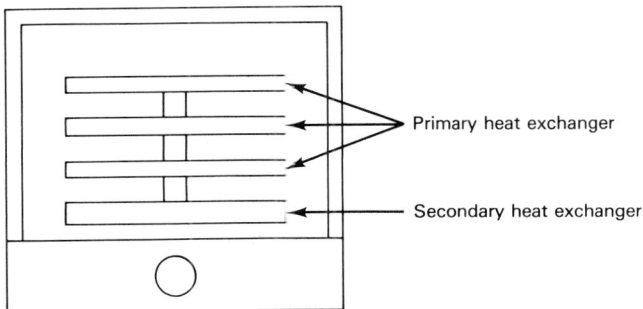

Figure 1.2 Additional heat exchanger for increased efficiency.

Primary Heat Exchangers

The primary heat exchangers are basically the same as those used on standard gas furnaces. They generally include the combustion zone and primary heat transfer surfaces. The heat exchangers used in condensing gas furnaces are made of a type of metal that has a high resistance to corrosion. This is necessary because of the very corrosive nature of the condensed products of combustion.

Condensate Drain

The secondary heat exchangers are equipped with a condensate drain connection that allows the condensate to be drained into the city drain system. This corrosive condensate should not be allowed to drain onto the ground or other places where damage can be caused. Plastic pipe is normally used as the drain line because of the corrosive nature of the condensate. Be sure to follow the proper procedure for installing the drain. This will be discussed in each manufacturer's section of this book and those recommendations should be followed unless there is a conflict with a local ordinance, in which case the local ordinance should be used.

Vent Pipe

Plastic pipe of the proper type, such as schedule-40 PVC, PVC-DWV, SDR-21 PVC, SDR-26 PVC, or ABS-DWV pipe, or any other recommended type, is used. The size is determined by the manufacturer of the furnace. Plastic pipe may be used because the flue gas temperature has been reduced below the safe operating temperature of the material. The size is generally smaller than the type B vent for the same size lower efficiency furnace because the water vapor has been removed and the reduced temperature causes a reduction in the volumetric area of the flue gases. The installation of the vent pipe should conform to the code that is in force at the time of installation at the job location.

Combustion Air

The combustion air is supplied to high-efficiency gas furnaces by a motor-driven blower. There are two types of combustion systems in use: forced draft and induced draft.

   Forced draft system.   The forced draft system forces the air into the combustion zone under pressure (see Figure 1.3). In this type of system the

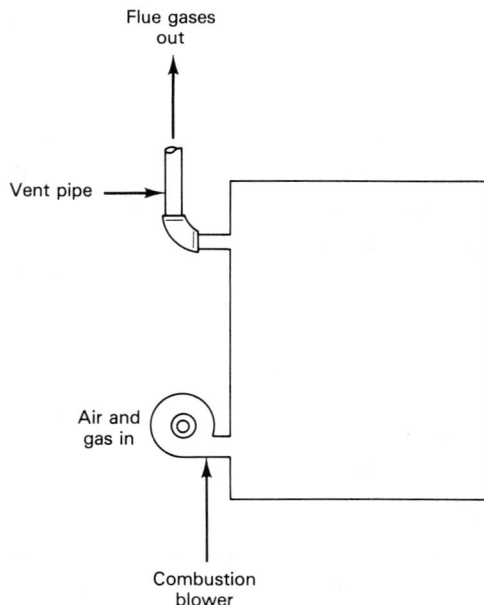

Figure 1.3 Forced draft heat exchanger.

combustion zone and the vent system have a slight pressure inside. The combustion air adjustments must be made according to the manufacturer's specifications for the furnace in question. The blower supplies all of the air needed for proper combustion and provides the force required to move the products of combustion through the vent system and into the atmosphere. Anytime that a leak occurs anywhere in the system the combustion process will be upset and proper, if any, operation will not be possible.

Induced draft system.   The induced draft system draws, or pulls, the air into the combustion zone through the heat exchanger and discharges it into the vent system (see Figure 1.4). The combustion zone and heat exchanger are under a negative pressure. The vent system may have either atmospheric pressure or a slightly positive pressure inside it. The combustion air adjustments must be made according to the manufacturer's specifications for the furnace in question. The blower supplies all the air needed for proper and complete combustion and provides the force required to move the products of combustion through the vent system and into the atmosphere. Anytime a leak occurs anywhere in the system the combustion process will be upset and proper, if any, operation will not be possible.

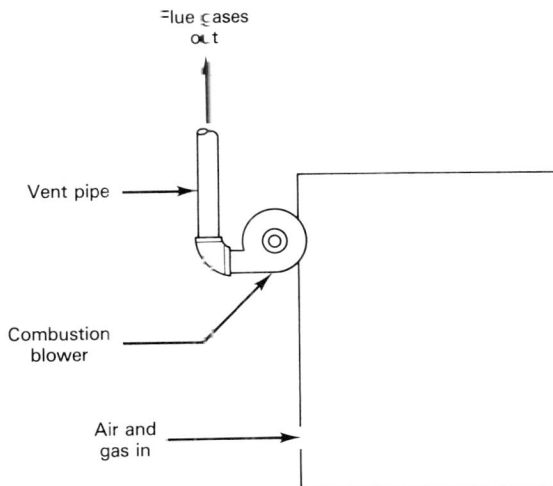

Figure 1.4 Induced draft heat exchanger.

Combustion and Ventilation Air Requirements

The flow of combustion and ventilation air to the furnace must not be restricted in any way. The air openings placed in the furnace cabinet during construction must be kept free of any obstructions that would in any way restrict the flow of air to the furnace. Such obstructions would affect the efficiency as well as the safe operation of the furnace. Furnaces must have a sufficient quantity of air for proper and safe operation and performance. This must be kept in mind during installation and service procedures.

For complete combustion and ventilation, the furnace requires approximately 20 ft³ of air for every 1000 Btu of gas burned. Thus, for each 1000 Btu of gas consumed, a total of 20 ft³ of air must be supplied to the furnace room.

> **Warning:** The air used for combustion and ventilation purposes must not come from a source containing corrosive contaminants. Therefore, when installing a condensing gas furnace in a commercial building, a workshop, or a beauty shop it may be necessary to provide outside air for combustion and ventilation purposes.

The combustion air must be free of all acid-forming chemicals, such as fluorine, chlorine, and sulfur. These elements are generally found in aerosol sprays, detergents, bleaches, cleaning solvents, air fresheners, paint and varnish removers, refrigerants, and many other commercial and household cleaning products. When the vapors from these products are burned in a gas flame, acid compounds are formed. These acid compounds then increase the dew point temperature of the flue gas products and are also highly corrosive after they are condensed.

Air for Combustion and Ventilation: (Extracted from the National Fuel Gas Code ANSI Z223.1–1980).

5.3.1 *General:*

    a. The provisions of 5.3 apply to gas utilization equipment installed in buildings and which require air for combustion, ventilation and dilution of flue gases from within the building. They do not apply to (1) direct vent equipment which is constructed and installed so that all air for combustion is obtained from the outside atmosphere and all flue gases are discharged to the outside atmosphere, or (2) enclosed furnaces which incorporate an integral total enclosure and use only outside air for combustion and dilution of flue gases.

    b. Equipment shall be installed in a location in which the facilities for ventilation permit satisfactory combustion of gas, proper venting and the maintenance of ambient temperature at safe limits under normal conditions of use. The equipment shall be located so as not to interfere with the proper circulation of air. When normal infiltration does not provide the necessary air, outside air shall be introduced.

    c. In addition to air needed for combustion, process air shall be provided as required for: cooling of equipment or material, controlling dew point, heating, drying, oxidation or dilution, safety exhaust, odor control, and air for compressors.

   d. In addition to air needed for combustion, air shall be supplied for ventilation, including all air required for comfort and proper working conditions for all personnel.

   e. While all forms of building construction cannot be covered in detail, air for combustion, ventilation and dilution of flue gases for gas utilization equipment vented by natural draft normally may be obtained by application of one of the methods covered in 5.3.3 and 5.3.4.

5.3.2 *Equipment Located in Unconfined Spaces:* In unconfined spaces in buildings, infiltration normally is adequate to provide air for combustion, ventilation, and dilution of flue gases.

5.3.3 *Equipment Located in Confined Spaces:* For equipment located in confined spaces use the following requirements:

   a. All air from inside the building: The confined space shall be provided with two permanent openings communicating directly with an additional room(s) of sufficient volume so that the combined volume of all spaces meets the criteria for an unconfined space. The total input of all gas utilization equipment installed in the combined space shall be considered in making this determination. Each opening shall have a minimum free area of 1 square inch per 1000 Btu per hour of the total input rating of all gas utilization equipment in the confined space, but not less than 100 square inches. One opening shall be within 12 inches of the top and one within 12 inches of the bottom of the enclosure.

   b. All air from outdoors: The confined space shall be provided with two permanent openings, one commencing within 12 inches of the top and one commencing within 12 inches of the bottom of the enclosure. The openings shall communicate directly, or by ducts, with the outdoors or spaces (crawl or attic) that freely communicate with the outdoors.

     1. When directly communicating with the outdoors, each opening shall have a minimum free area of 1 square inch per 4000 Btu per hour of total input of all equipment in the enclosure.

     2. When communicating with the outdoors through vertical ducts, each opening shall have a minimum free area of 1 square inch per 4000 Btu per hour of the total input rating of all equipment in the enclosure.

     3. When communicating with the outdoors through horizontal ducts, each opening shall have a minimum free area of 1 square inch per 2000 Btu per hour of total input rating of all equipment in the enclosure.

     4. When ducts are used, they shall be of the same cross-sectional area as the free area of the openings to which they connect. The minimum dimension of rectangular air ducts shall be not less than 3 inches.

5.3.4 *Specially Engineered Installations:* The air requirements of 5.3.3 shall not necessarily govern when special engineering, approved by the authority having jurisdiction, provides an adequate supply of air for combustion, ventilation, and dilution of flue gases.

5.3.5 *Louvers and Grills:* In calculating free area in 5.3.3, consideration shall be given to the blocking effect of louvers, grills or screens protecting openings. Screens used shall not be smaller than 1/4 inch mesh. If the free area through a design of louver or grill is known, it should be used in calculating the size opening required to provide the free area specified. If the design and free area is not known, it may be assumed that wood louvers will have 20–25 percent free area and metal louvers and grills 60–75 percent free area. Louvers and grills shall be fixed in the open position or interlocked with the equipment so that they are opened automatically during equipment operation.

5.3.6 *Special Considerations Created by Mechanical Exhausting or Fireplaces:* Operation of exhaust fans, ventilation systems, power burners, induced draft systems, clothes dryers, or fireplaces may create conditions requiring special attention to avoid unsatisfactory operation of installed gas utilization equipment. In especially cold climates, it is possible that the use of openings of the sizes specified above in outside walls may result in over ventilation or excessive cooling of the utility room or other confined space furnace locations. Under certain conditions, this might introduce the hazard of freezing water lines or water heater storage tanks. In this case, a supply air duct should be used to heat this space.

Air openings in the casing front, return grill, and warm air diffusers or registers must be kept free of obstructions restricting air flow.

EXAMPLE 1:   Furnace Located in Unconfined Space

An unconfined space (such as an open basement) must have a minimum volume of 50 ft^3 per 1000 Btuh of the total of all the appliances in the area, only if there are no doors between the rooms.

Figure 1.5 shows the required minimum area in square feet to qualify as an unconfined space for different Btuh input ratings. The table is based on rooms with 8-ft ceiling heights. The table also shows the required round duct size to provide the necessary outside air and the maximum Btuh input that the duct size will handle.

If the area is an unconfined space, provide an opening or openings having a total free area of 1 in.^{2} per 4000 Btuh of the total of all appliances. The required duct size is shown in Table 1.1, or refer to the required size in square inches in the column (4000) shown in Table 1.2.

Figure 1.5 shows a typical duct going outdoors. The duct may also go into a ventilated crawl space or attic for a ground floor installation.

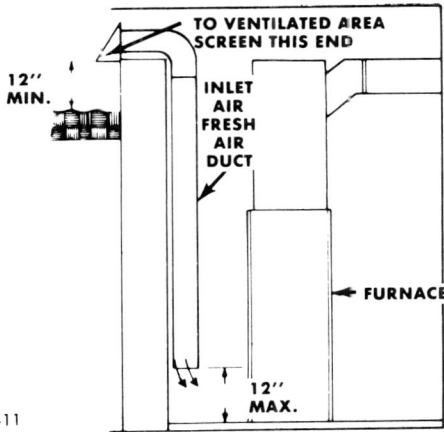

21-10-11

Figure 1.5 Fresh air duct (basement). (Courtesy of Heil-Quaker Corporation.)

TABLE 1.1 MINIMUM AREA (FT^2) (COURTESY OF HEIL-QUAKER CORPORATION)

4000 Btuh per square inch round duct size	Max. Btuh input	Unconfined space min. area in sq. ft. 8′ ceiling height
4″	50,000	312
5″	78,500	490
5-1/8″	80,000	500
5-7/8″	105,000	656
6″	114,000	712
7″	155,000	968
8″	200,000	1250

TABLE 1.2 FREE AREA IN SQUARE INCHES EACH OPENING (FURNACE ONLY) (COURTESY OF HEIL-QUAKER CORPORATION)

Furnace maximum Btuh input rating	One sq. inch of opening for each (xx) BTU per hour of the input rating	
	(2000)	(4000)
	Square inches of each opening	
40,000	20	10
50,000	25	13
75,000	38	19
100,000	50	25
125,000	63	32

1. The outside air intake must be at least 1 ft above ground level and be protected from obstructions.
2. Protect the air intake with a screen not less than 1/4-in. mesh.
3. The duct must terminate at a point not more than 1 ft above the floor.

> **Caution:** Do not connect the air intake duct to the furnace or terminate near an air inlet grill. Extreme cold air may cause the condensate to freeze during the off cycle and damage the furnace.

EXAMPLE 2:   Furnace Located in Confined Space

If the furnace is installed in a room or area considered to be a confined space, it must be provided with free air for proper combustion and ventilation of flue gases.

Provide two permanent openings, one within 12 in. of the furnace top, one within 12 in. of the bottom of the room connecting directly, or by using ducts, with the outdoors or areas open to the outdoors. A minimum of one inlet and one outlet is required using any of the combinations shown in Figures 1.6–1.8.

If the opening connects directly to, or with vertical ducts, the free area of each opening must be at least 1 in.^{2} per 4000 Btuh combined input of all appliances in the confined space.

If horizontal ducts are used, the free area of each opening must be at least 1 in.^{2} per 2000 Btuh combined input of all the appliances in the confined space. For example, the furnace is rated at 105 000 Btu per hour. The water heater is rated at 25 000 Btu per hour. The total is 130 000 Btu per hour. Two grills are needed each with 33 in.^{2} of free opening, unless connected by horizontal ducts, which would require each grill or opening to have a free area of 65 in.^{2}.

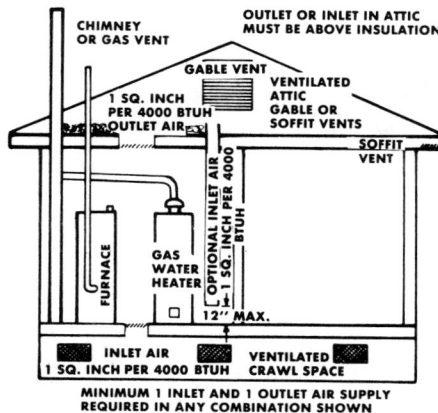

Figure 1.6 Air from attic and crawl space. (Courtesy of Heil-Quaker Corporation.)

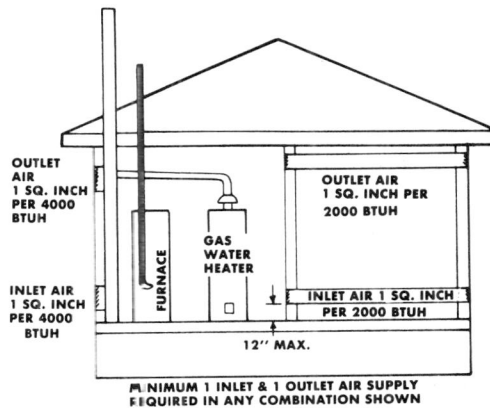

Figure 1.7 Outside air using horizontal inlet and outlet. (Courtesy of Heil-Quaker Corporation.)

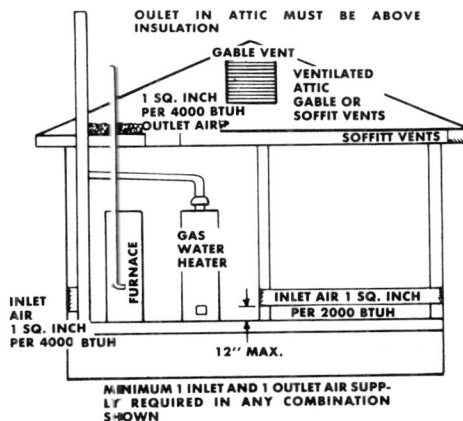

Figure 1.8 Outside air using horizontal inlet/attic outlet. (Courtesy of Heil-Quaker Corporation.)

VENT PIPE INSTALLATION

Warning: Danger of property damage, bodily injury or death. Proper vent pipe installation is critical to the safe operation of the furnace. Carefully read and follow these instructions and those that apply to the furnace in question.

Condensing gas furnaces remove both sensible and latent heat from the combustion flue gases. Removal of the latent heat results in the condensation of flue gas water vapor. The condensed water vapor drains from the secondary heat exchanger into the combustion blower and out of the unit into a PVC drain trap (see Figures 1.9 and 1.10).

The furnace must be vented to the outdoors using the proper connections

TYPICAL STANDARD CONNECTION

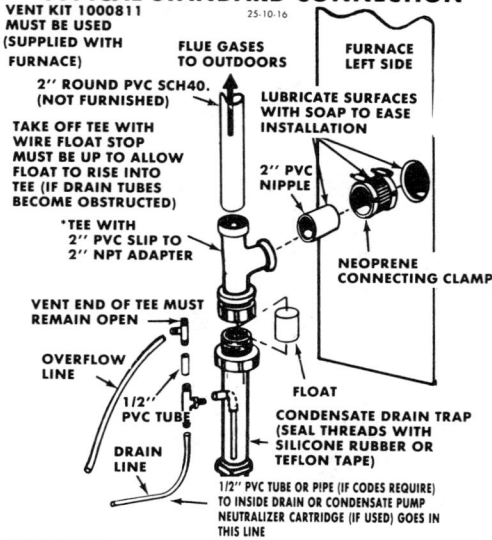

VENT KIT 1000811
MUST BE USED
(SUPPLIED WITH
FURNACE)

25-10-16

FLUE GASES
TO OUTDOORS

FURNACE
LEFT SIDE

2'' ROUND PVC SCH40.
(NOT FURNISHED)

LUBRICATE SURFACES
WITH SOAP TO EASE
INSTALLATION

TAKE OFF TEE WITH
WIRE FLOAT STOP
MUST BE UP TO ALLOW
FLOAT TO RISE INTO
TEE (IF DRAIN TUBES
BECOME OBSTRUCTED)

2'' PVC
NIPPLE

*TEE WITH
2'' PVC SLIP TO
2'' NPT ADAPTER

NEOPRENE
CONNECTING CLAMP

VENT END OF TEE MUST
REMAIN OPEN

OVERFLOW
LINE

1/2''
PVC TUBE

FLOAT

DRAIN
LINE

CONDENSATE DRAIN TRAP
(SEAL THREADS WITH
SILICONE RUBBER OR
TEFLON TAPE)

1/2'' PVC TUBE OR PIPE (IF CODES REQUIRE)
TO INSIDE DRAIN OR CONDENSATE PUMP
NEUTRALIZER CARTRIDGE (IF USED) GOES IN
THIS LINE

* DO NOT COUNT TEE OR TERMINATION ELBOW (HORZ. VENT) WHEN
DETERMINING VENT LENGTH

TYPICAL ALTERNATE CONNECTIONS

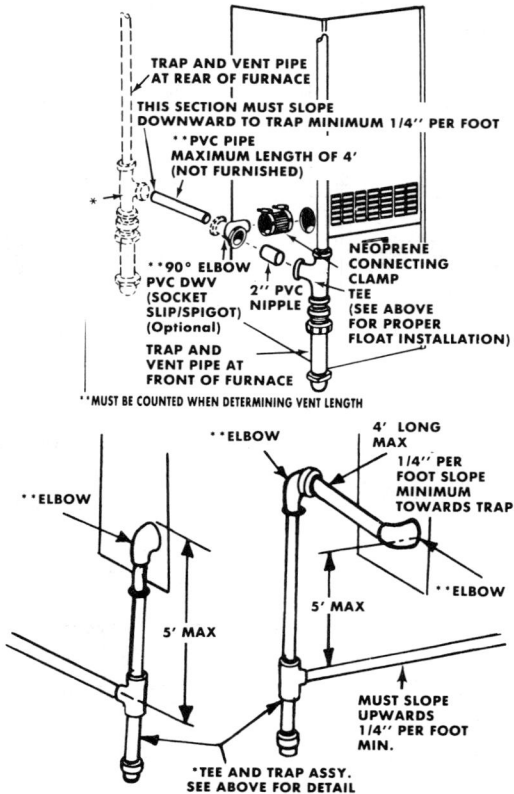

TRAP AND VENT PIPE
AT REAR OF FURNACE

THIS SECTION MUST SLOPE
DOWNWARD TO TRAP MINIMUM 1/4'' PER FOOT

**PVC PIPE
MAXIMUM LENGTH OF 4'
(NOT FURNISHED)

*

NEOPRENE
CONNECTING
CLAMP

**90° ELBOW
PVC DWV
(SOCKET
SLIP/SPIGOT)
(Optional)

2'' PVC
NIPPLE

TEE
(SEE ABOVE
FOR PROPER
FLOAT INSTALLATION)

TRAP AND
VENT PIPE AT
FRONT OF FURNACE

**MUST BE COUNTED WHEN DETERMINING VENT LENGTH

**ELBOW

4' LONG
MAX

1/4'' PER
FOOT SLOPE
MINIMUM
TOWARDS TRAP

**ELBOW

**ELBOW

5' MAX

5' MAX

5' MAX

MUST SLOPE
UPWARDS
1/4'' PER FOOT
MIN.

*TEE AND TRAP ASSY.
SEE ABOVE FOR DETAIL

Figure 1.9 Vent trap and furnace connection. (Courtesy of Heil-Quaker
Corporation.)

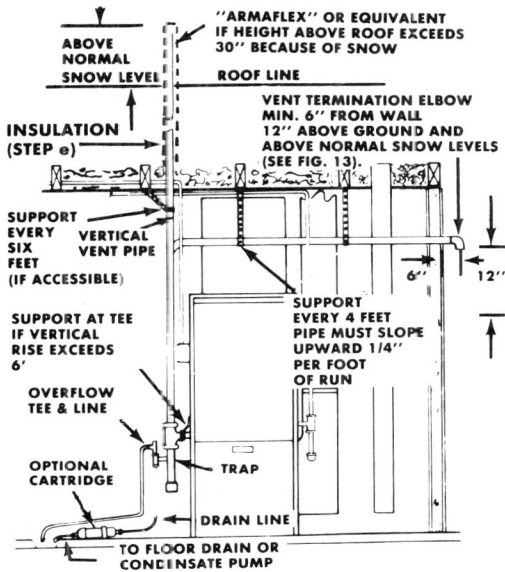

Figure 1.10 Vent installation. (Courtesy of Heil-Quaker Corporation.)

and the proper size vent pipe, which is made from the required type of material in accordance to the local codes and ordinances.

When a substitute piping is used it must be connected to the furnace with the proper fittings for the type of pipe being used. All joints, fittings, and so on must be cemented, sealed, or mechanically connected to prevent leakage of the flue gases.

All of the instructions, guidelines, and limitations outlined in this section for PVC piping must be followed unless they are in conflict with the type of material being used, local codes, or the requirements of the furnace manufacturer.

The vent must be installed in compliance with Part 7, Venting of Equipment, of the National Fuel Code NFPA 54/ANSI Z 223.1, 1984, local codes or ordinances, these instructions, and good trade practices.

Each vent must serve only one furnace. Do not connect to an existing vent or chimney.

Vertical venting is preferred because there will be some moisture in the flue gases that may condense in the vent pipe (See Special Instructions for Horizontal Vents).

Instructions

The vent must exit the furnace at the designated point. If necessary a drain trap assembly must be used, such as when a side vent exit is used, to provide the necessary 5-in. water column (wc) against the vent pressure. The drain trap must be constructed with the proper parts. Make sure that all parts fit properly and are correctly oriented before beginning any solvent cementing. Observe the following guidelines and limitations when constructing the vent assembly.

a. The vent diameter must not be reduced.
b. The drain trap assembly, when used, must be within 4 ft horizontally and 5 ft vertically (lower only) of the furnace vent connector. Some typical examples are shown in Figure 1.9. All vent piping from the furnace to the trap must slope downward a minimum of 1/4 in. per foot of run. All vent piping from the trap to the vent termination must slope upward a minimum of 1/4 in. per foot of run.

   The drain trap assembly may not be installed in any unconditioned space if there is any chance of condensate freezing inside the trap or drain lines.

   The drain trap must be reasonably accessible so that homeowners can check it.

   > **Note**: Any elbows used to change from a vertical to a horizontal run should be of the type DWV to provide the correct slope in the horizontal run. If other type elbows are used, then two 45° elbows should be used, in place of one 90°, with the elbows slightly misaligned to provide a slope in the horizontal runs.

c. All horizontal pipe runs must be supported at least every 4 ft with metal pipe strapping. No sags or dips are permitted.
d. All vertical vent pipe runs must be supported every 6 ft where accessible.
e. The vent pipe must be insulated if there is any chance of condensate freezing inside the pipe. This can occur if the vent pipe passes through an unconditioned space, such as an attic, crawl space, uninsulated chase, or a masonry chimney. It can also occur where the vent terminates above the roof or if an exterior vertical riser is used to get above snow levels (see Figure 1.11). The local climatic conditions and the vent length must be considered.

   If the vent height above the roof exceeds 30 in. because of snow accumulations it must be insulated.

   Insulation: For exterior or interior use: Armaflex or equivalent closed-cell foam insulation should be used. The recommended thickness is 1 in., or multiple layers if required for extreme climate conditions.

   For Exterior Uses Only: Fiberglass or equivalent with a vapor barrier is recommended. The recommended R value of 7 up to 10 ft is R-11 if the exposure exceeds 10 ft.

f. If it is necessary to insulate the vent pipe and a chimney is used as a

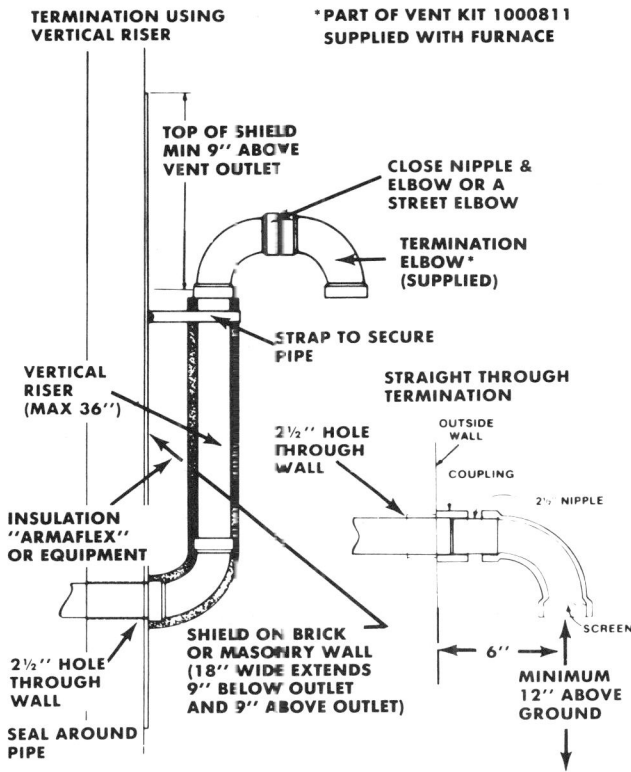

TERMINATION USING VERTICAL RISER

*PART OF VENT KIT 1000811 SUPPLIED WITH FURNACE

TOP OF SHIELD MIN 9" ABOVE VENT OUTLET

CLOSE NIPPLE & ELBOW OR A STREET ELBOW

TERMINATION ELBOW* (SUPPLIED)

STRAP TO SECURE PIPE

VERTICAL RISER (MAX 36")

STRAIGHT THROUGH TERMINATION

OUTSIDE WALL

2½" HOLE THROUGH WALL

COUPLING

2½ NIPPLE

INSULATION "ARMAFLEX" OR EQUIPMENT

SCREEN

SHIELD ON BRICK OR MASONRY WALL (18" WIDE EXTENDS 9" BELOW OUTLET AND 9" ABOVE OUTLET)

2½" HOLE THROUGH WALL

←— 6" —→

MINIMUM 12" ABOVE GROUND

SEAL AROUND PIPE

Figure 1.11 Vent termination. (Courtesy of Heil-Quaker Corporation.)

chase, the top of the chimney must be sealed flush, or crowned up, so that only the vent pipe protrudes.

g. When the vent height above the roof exceeds 30 in., or if an exterior vertical riser is used on a horizontal vent to get above the snow levels, the exterior portion must be insulated. Use only moisture-resistant insulation, such as Armaflex.

h. The maximum vent length is 60 total equivalent feet with each 45° elbow (maximum of 8) counting as 2 1/2 ft and each 90° elbow (maximum of 4) counting as 4 ft. Do not count the take off tee (Figure 1.9) or the vent termination elbow on a horizontal vent. Do not use the termination elbow on a vertical vent through the roof. *Example*: A 40-ft vent pipe with four 90° elbows, (20 ft) equals 60 equivalent feet; or 40 ft of vent pipe with two 45° elbows (5 ft) and three 90° elbows (15 ft) equals 60 equivalent feet.

i. The minimum vent length is 5 ft.

j. Do not install the vent pipe in the same chase with a vent from another gas or other fuel-burning appliance.

    k. Do not install the vent pipe within 6 in. of a vent pipe from another gas or other fuel-burning appliance.

    l. The vent pipe can run in the same chase or adjacent to supply or vent pipe for water supply or waste plumbing.

Connecting to the Furnace

The vent pipe must be fastened to the furnace by using the manufacturer's recommended procedure. Do not attempt to cement the pipe to the blower housing.

If a vertical rise exceeds 6 ft, secure the tee and trap assembly to the furnace with a mounting strap to remove excessive weight from the clamp connection.

An optional 90° elbow (PVC DWV socket × slip/spigot) may be used and fastened to the combination blower outlet coupling using a 2 × 2-in. PVC nipple. See Figure 1.9 for the drain trap assembly.

Joining Pipe and Fittings

All pipe, fittings, solvent cement, and procedures must conform to the American National Standard Institute and the American Society for Testing and Materials (ANSI/ASTM) standards.

Pipe and fittings. ASTM D 1785, D 2466 & D 2665 PVC primer and solvent cement, ASTM D 2564 Procedure for Cementing Joints, Reference ASTM D 2855.

Warning: Danger of fire or bodily injury. PVC solvent cements and primers are highly flammable. Provide adequate ventilation and do not assemble near a heat source or open flame. Do not smoke. Avoid skin or eye contact. Observe all cautions and warnings printed on material containers.

All joints in the PVC vent must be properly sealed using the following material and procedure.

Caution: For proper installation:
Do not use solvent cement that has become curdled, lumpy, or thickened.
Do not thin. Observe shelf precautions printed on the containers.
For applications below 32 °F use only a low-temperature-type solvent.

PVC cleaner-primer and PVC medium body solvent cement.

1. Cut the pipe and square the end; remove ragged edges and burrs. Chamfer the end of the pipe; then clean the fitting socket and pipe joint area of all dirt, grease, or moisture.
2. After checking the pipe and socket for proper fit, wipe the socket and pipe with cleaner-primer. Apply a liberal coat of primer to the inside surface of the socket and the outside of the pipe. Do not allow the primer to dry before applying the cement.
3. Apply a thin coat of cement evenly in the socket. Then quickly apply a heavy coat of cement to the pipe end and insert the pipe into the fitting with a slight twisting motion until it bottoms out.

> **Note**: The cement must be fluid; if not, recoat.

4. Hold the pipe in the fitting for 30 seconds to prevent the tapered socket from pushing the pipe out of the fitting.
5. Wipe all excess cement from the joint with a rag. Allow 15 minutes before handling. Cure time varies according to the fit, temperature, and humidity.

> **Note**: Stir the solvent cement frequently while using it. Use a natural bristle brush or the dauber supplied with the can. The proper brush size is 1 in.

Condensate Drain/Neutralizer

The drain line and the overflow line can be 1/2 in. PVC flex tube or schedule 40 with a disconnect union so the trap can be removed. The trap assembly provides 5 in. water column so no additional trap is required. Drains must terminate at an outside drain.

> **Warning**: Do not run an outside drain. Freezing of the condensate could cause property damage.

If a condensate pump is used, or if local codes require one, install a condensate neutralizer cartridge in the drain line. Be sure to install the cartridge in a horizontal position only. Install an overflow line if routing to a floor drain or a sump pump (see Figure 1.9).

If no inside floor drain is available, a condensate pump or sump pump must be used. The condensate neutralizer must be used with either type of pump.

The condensate pump must have an auxiliary safety switch to prevent operation of the furnace and resulting overflow of condensate in the event of pump failure. The safety switch must be wired through the R circuit *only* (low voltage) to provide operation in either of the heating or cooling modes.

Horizontal Vents

The furnace may be vented horizontally through an outside wall, using all of the applicable instructions under Vent Pipe Installation with these additional requirements. The requirements and limitations for horizontal venting are very strict. *All horizontal vent installations must be made in accordance with these instructions.*

Vent location.   The vent location must meet the requirements listed in the following instructions or applicable codes, whichever specifies the most clearance or strictest limitations.

> **Warning**: The combustion products and moisture in the flue gases may condense as they leave the terminal elbow. The condensate may freeze on the exterior wall, under the eaves, and on surrounding objects. Some discoloration to the exterior of the building may occur.

Location requirements.   The vent must be installed with the following clearances and requirements (see Figure 1.12).

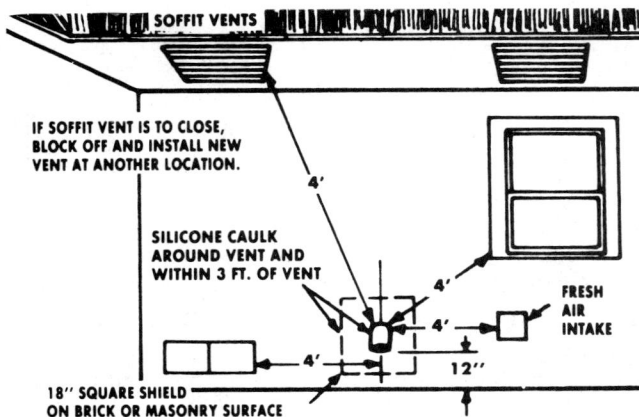

Figure 1.12 Minimum clearances. (Courtesy of Heil-Quaker Corporation.)

a. Twelve inches above ground level, above normal snow levels (when practicable), and 6 inches out from the wall.

> **Note:** Ice or snow may cause the furnace to shut down if the vent becomes obstructed. If required use a vertical riser or shield vent to prevent blockage from drifting snow (see Figure 1.13).

b. Not above any walkway or area that may create a hazard or nuisance or be detrimental to the operation of other equipment.
c. Four feet from and not above or below any door, window, gravity inlet, or forced air inlet for the building.
d. At least 4 ft from any soffit or under an eave vent.
e. Do not vent under any kind of patio or deck.
f. Locate the vent on the side of the building away from prevailing winter winds when practical but taking into consideration other limitations to determine the best overall location. If installed on a side with prevailing

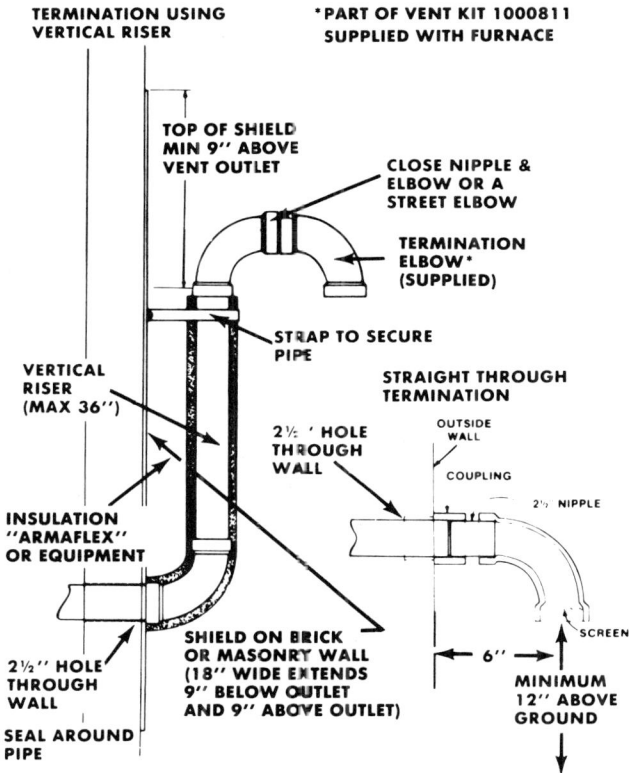

Figure 1.13 Typical vent piping. (Courtesy of Heil-Quaker Corporation.)

winds consider the possible effects of moisture damage from freezing on the walls or overhangs (under eaves) and use protective measures such as shielding (step g) and/or sealing cracks, seams, and joints (step i) but extend the area of sealing to a minimum of 6 ft.

g. On brick or masonry surfaces, use a rust-resistant shield (18 in.^{2}) behind the vent. If a vertical rise is used the shield must extend 9 in. above and 9 in. below, as shown in Figure 1.13. The shield can be wood, plastic, sheet metal, etc.

h. Do not locate the vent too close to shrubs, because condensate may stunt or kill them.

i. Caulk all cracks, seams, and joints within 3 ft of the vent.

Vent Termination.   The vent termination elbow must be installed as shown in Figure 1.13.

a. Cut a 2 1/2-in.-diameter hole through the exterior wall. Do not make the hole oversized or it will be necessary to add a sheet metal or plywood plate on the outside with the correct size hole in it. Check the hole size by making sure it is smaller than the coupling or elbow that will be installed on the outside. The coupling or elbow must prevent the pipe from being pushed back through the wall.

b. Extend the vent pipe through the wall 3/4 to 1 in. and seal the area between the pipe and the wall.

Straight Through Termination (No Vertical Riser):

c. Install the coupling, 2 1/2-in.-long nipple, and termination elbow as shown in Figure 1.13.

Termination Using Exterior Riser:

d. Install the elbows and vent pipe (maximum of 36 in. long) to form a riser, as shown in Figure 1.13.

e. Secure the vent pipe to the wall with a galvanized strap or other rust-resistant material to restrain the pipe from moving.

f. Insulate the pipe with Armaflex or equivalent moisture-resistant, closed-cell foam insulation.

Note: If situations require that the pipe be run on the exterior of the wall to reach a suitable termination location it must be properly insulated. It must be boxed in and sealed against moisture if fiberglass insulation is used.

GAS SUPPLY AND PIPING

The American Gas Association (AGA) rating plate is stamped with the model number, type of gas, and gas input rating.

> **Warning** Danger of property damage, bodily injury, or death. Make sure that the furnace is equipped to operate on the type of gas available. Models designated as natural gas are to be used with natural gas only.

Furnaces designated for use with liquified petroleum (LP) gas have orifices sized for commercially pure propane gas. They must not be used with butane or a mixture of butane and propane unless properly sized orifices are installed by a licensed LP installer.

Gas Supply

The recommended gas supply pressures are 7 in. water column pressure for natural gas and 11 in. water column pressure for LP gas. A maximum gas supply pressure of 14 in. water column should not be exceeded on either gas. A minimum gas supply pressure of 4 1/2 in. water column for natural gas and 11 in. water column for LP is required for the purpose of input adjustment and it should not be allowed to vary downward, because this will decrease the input to the unit.

The gas input to the burners must not exceed the rated input shown on the rating plate. On natural gas the manifold pressure should be 3 1/2 in. water column. The manifold pressure should be 10 in. water column for LP gas. For operation at altitudes above 2000 ft an orifice change or a manifold pressure adjustment may be required to suit the gas being supplied. Check with the gas supplier.

For elevations over 2000 feet the furnace should be derated based on the standard input rating. Do not base on the alternate input rating.

Orifice Sizes

Make certain that the unit is equipped with the correct main burner and pilot burner orifices.

Factory-sized orifices for natural and LP gas are generally listed in the manufacturer's literature.

Gas Piping

The gas pipe supplying the furnace must be properly sized to handle the combined appliance loads or it must run directly from the gas meter (natural gas) or LP gas second-stage regulator and supply only the furnace. It must be the correct size for the length of run and furnace rating. The length of pipe or tubing should be measured from the gas meter for natural gas or the LP gas second-stage regulator, which is usually just outside the building wall.

Determine the minimum pipe size from the proper table, basing the length of the run from the main line, gas meter, or source to the furnace (see Tables 1.3 and 1.4).

TABLE 1.3 GAS PIPE SIZES/CAPACITY, NATURAL GAS
(COURTESY OF HEIL-QUAKER CORPORATION)

Length of pipe–ft	Capacity-Btuh per hour input		
	1/2″	3/4″	1″
20′	92,000	190,000	350,000
40	63,000	130,000	245,000
60	50,000	105,000	195,000

TABLE 1.4 GAS TUBING AND PIPE SIZES, LP GAS (COURTESY OF HEIL-QUAKER CORPORATION)

Length in feet	Capacity-Btuh per hour input			
	1/2″**	3/4″**	1/2″	3/4″
20′	62,000	216,000	189,000	393,000
40′	41,000	145,000	129,000	267,000
60′	35,000	121,000	103,000	217,000

*Copper tubing for gas supply must comply with limitations in National Fuel Gas Code, reference "2.6.3 Metallic Tubing"
**Outside diameter.

Note: Use the correct size of pipe. Piping that is too small will not allow enough gas to reach the furnace and will reduce the heat output of the furnace.

Check the gas line installation for compliance with the local codes.

Connecting the Gas Piping: Refer to Figure 1.14 for the general layout at the furnace. It shows the basic fittings you will need. The following rules apply:

21-10-29

RIGHT SIDE ENTRY

MANUAL VALVE

8" NIPPLE

GAS VALVE

1/2 UNION

ELBOW

3-1/2" NIPPLE

STREET ELBOW

DO NOT SECURE OR SUPPORT CONNECTOR TO FURNACE

USE ELBOW OR FITTING TO MATCH WITH CONNECTOR

APPROVED GAS CONNECTOR (SEE PIPING PAGE 12)

1/2" PIPE AND FITTINGS TO GAS VALVE (A UNION MAY BE REQUIRED DEPENDING ON TYPE OF CONNECTOR)

GAS VALVE

1/2 UNION

NIPPLE

MANUAL VALVE 4' ABOVE FLOOR

1/8" NPT PLUG (TEST GAUGE)

DRIP LEG

1/2" x 4"

1/2" PIPE CAP

LEFT SIDE ENTRY

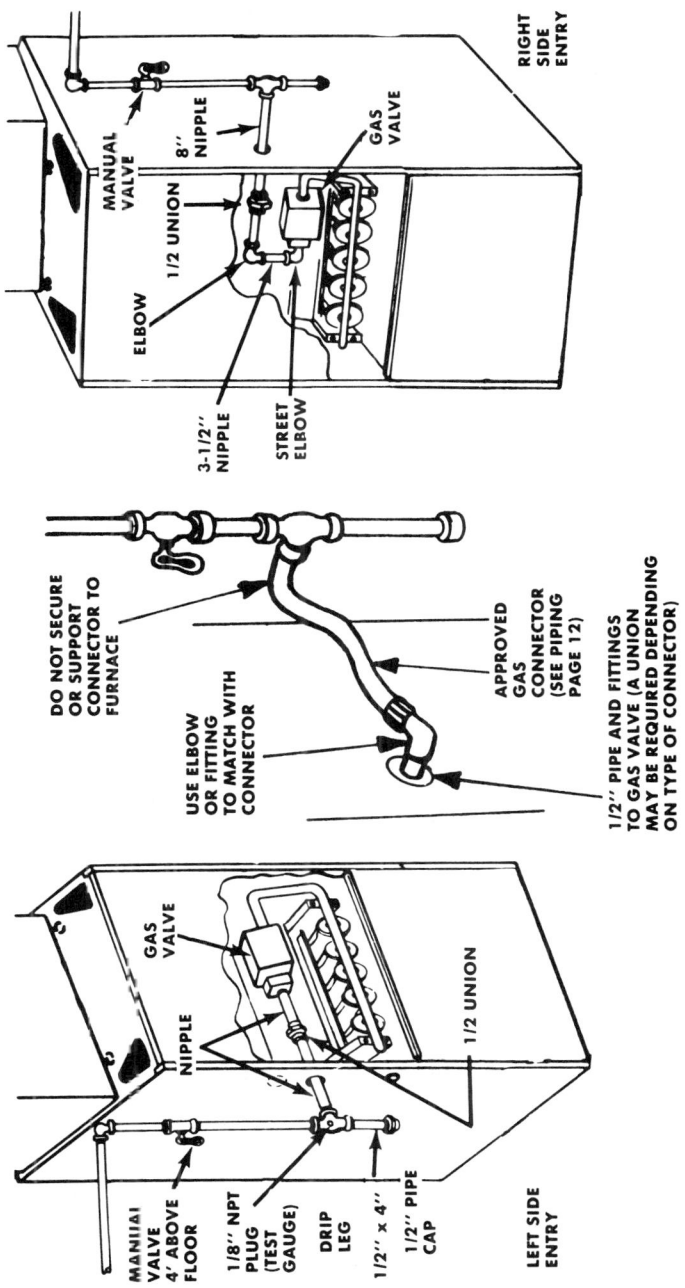

Figure 1.14 Proper piping practice. (Courtesy of Heil-Quaker Corporation.)

23

1. Use black iron or steel pipe and fittings or other piping that is approved by local codes.
    a. If a gas connector is used it must be acceptable to the local authority. The connector may not be used inside the furnace or be secured or supported by the furnace or the ductwork. The connectors should comply with one of the following standards or a superseding standard:
    ANSI Z 21.24a, 1983, Metal Connectors for Gas Appliances
    ANSI Z 21.45b, 1983, Flexible Connectors of Other Than All-Metal Construction for Gas Appliances
2. Use pipe joint compound on the male threads only. The pipe joint compound must be resistant to the action of the LP gases (see Figure 1.15).

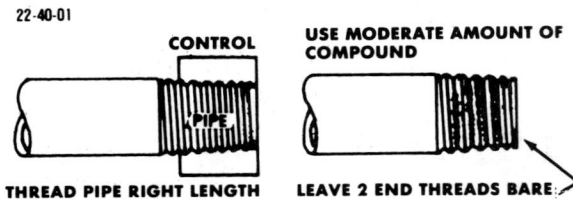

Figure 1.15 Application of pipe joint compound. (Courtesy of Heil-Quaker Corporation.)

3. Use ground joint unions.
4. Install a drip leg to trap dirt and moisture before it can enter the gas valve. The drip leg must a minimum of 3 in. long.
5. Use two pipe wrenches when making the connections to the valve to keep it from turning.
6. Provide a 1/8-in. National Pipe Thread (NPT) plug for a test gauge connection immediately upstream of the gas supply connection to the furnace.
7. Install a manual shutoff valve.
8. Tighten all joints securely.

    Additional LP Gas Requirements:
9. All connections made at the storage tank should be made by a licensed LP gas dealer.
10. An LP gas dealer should check all the lines and connections from the storage tank to the heating unit when the unit is connected to the storage tank.
11. Two-stage regulators should be used by the LP installer.
12. All gas piping should be checked out by the LP installer.

Checking the Gas Piping: Test all piping for leaks. When checking gas piping to the furnace, shut off the manual gas valve for the furnace. The gas pressure must not exceed 1/2 psig. If the gas piping is to be checked with pressure above 1/2 psig, the furnace and manual valve must be disconnected during the testing procedure. Apply soap suds (or a liquid detergent) to each joint. The formation of bubbles indicates a leak. Correct even a very small leak at once.

If the orifices were changed, make sure the pilot tube and burner orifices are checked for leakage.

> **Warning**: Danger of property damage, bodily injury, or death. Never use a match or open flame to test for leaks. Never exceed the specified pressures for testing. Higher pressures may damage the gas valve and cause overfiring which may result in heat exchanger failure. Liquifiec petroleum gas is heavier than air and it will settle in any low area, including open depressions, and will remain there unless the area is properly ventilated.

Never attempt to start up the unit before thoroughly ventilating the area.

ELECTRICAL WIRING

> **Warning**: Danger of bodily injury or death. Turn off the electric power at the fuse box or service panel before making any electrical connections. A grounded connection must be completed before making the line voltage connections. All line voltage connections must be made inside the furnace junction box. All electrical work must conform with the requirements of local codes and ordinances and the National Electrical Code ANSI/NFPA, No. 70, 1984 or current edition.

Grounding

A green wire pigtail is generally installed for the ground connection. Use an insulated copper conductor (No. 14 AWG) from the unit to a ground connection in the electric service panel or a properly driven and electrically grounded ground rod.

Electric Power Supply

The line voltage section is completely factory wired. It is only necessary to run No. 14 AWG hot, neutral, and ground wires from the power supply circuit (15 A) through a disconnect switch (if required by codes) to furnish power to the unit. Do not connect to existing lighting or other circuits.

Do not complete the line voltage connections until the unit is permanently grounded. All line voltage connections and the ground connection must be made with copper wire (see Figure 1.16).

Optional Equipment Wiring

All wiring (except the thermostat) from the furnace to the optional equipment, such as humidifiers or electronic air cleaners, or between optional equipment, must conform to the temperature limitations for Type T wire and be installed in accordance with the manufacturer's instructions supplied with the equipment.

Humidifier/Electronic Air Cleaner Wiring

The power connection for a humidifier or electronic air cleaner must be made through a sail switch, installed in the ductwork, if the furnace has a SPDT fan relay, with only three terminals.

If the manufacturer does not supply a sail switch, consult the place of purchase.

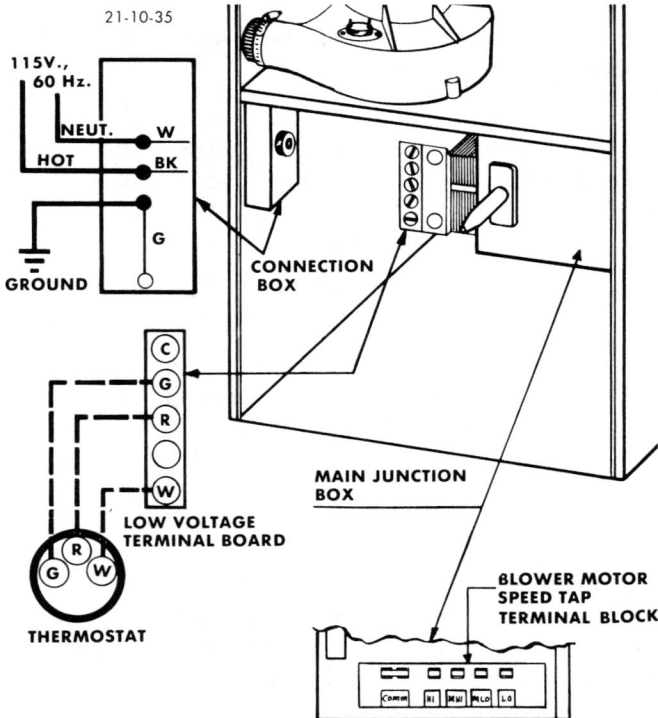

Figure 1.16 *Electrical connection.*
(Courtesy of Heil-Quaker Corporation.)

If the furnace has a DPDT fan relay with six terminals the power connections can be made to the furnace fan relay (see Figure 1.17).

With these connections and those shown in Figure 1.16 the humidifier will be powered when the furnace is fired and the circulating blower comes on. The electronic air cleaner will be powered anytime the circulating air blower is on, whether for heating, cooling, or just the fan on for air circulation.

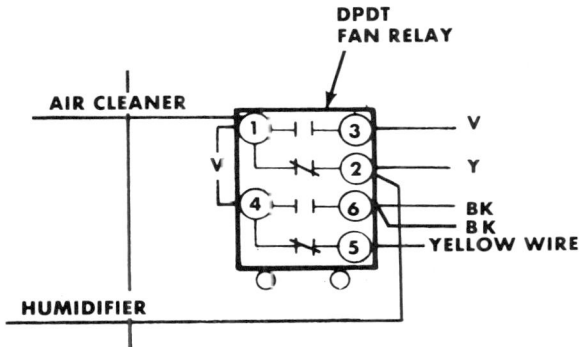

Figure 1.17 *Humidifier and air cleaner connections. (Courtesy of Heil-Quaker Corporation.)*

Blower Speeds

When it is necessary to change the blower speed, the manufacturer's specifications for the particular furnace in question should be consulted. This procedure will save time and money and perhaps prevent damage to the equipment.

Thermostat

The location of the thermostat has an important effect on the operation of the unit. Follow the instructions that are included with the thermostat for correct mounting and wiring.

Heat Anticipator: Set the heat anticipator in accordance with the thermostat instructions to the values shown on the gas valve or the manufacturer's data.

Thermostat Connection, Heating Only: Connect the two wires from the thermostat to terminals R and W on the transformer low-voltage terminal board. If the thermostat has a fan ON switch it will connect to terminal G (see Figure 1.16).

Adding Air Conditioning

1. Obtain a heating–cooling thermostat and 4-wire thermostat cable. Replace the existing thermostat and cable. Connect the wires to Y, W, G,

and R on the low-voltage terminal board to Y, W, G, and R on the thermostat.

2. The condensing unit will have a contactor in it. Connect its 24-V coil to terminals Y and C on the low-voltage terminal board. It may be necessary to consult the manufacturer's data for connections other than these.

3. Follow all of the instructions packed with the condensing unit and the evaporator coil.

The furnace fan relay will now change fan speeds automatically as the thermostat is switched to HEAT or COOL.

DUCTWORK AND FILTER

> **Warning**: Danger of bodily injury or death. The return air must not be drawn from inside the furnace closet or utility room. The return air duct must be sealed to the furnace casing.

Cool air from an evaporator coil passing over the heat exchanger may cause condensate to form inside the heat exchanger, causing failure of the heat exchanger.

The air distribution system should be designed and installed in conformation with manuals published by ASHRAE or other approved methods in conformance with local codes and good accepted trade practices.

When a furnace is installed so that the supply air ducts carry air circulated by the furnace to an area outside the space containing the furnace, the return air must also be handled by a duct or ducts sealed to the furnace casing and terminating outside the space containing the furnace. This is to prevent drawing possibly hazardous combustion products into the circulated air.

When air conditioning is installed with the furnace, the air conditioning cooling coil (evaporator) must be on the outlet side of the furnace or an evaporator and a blower can be separate from the furnace. This means the same duct would be used but the air will go around the furnace during the cooling cycle. With a separate blower and evaporator the dampers must seal properly for good air flow control. Chilled air going through the furnace could cause condensation and shorten the furnace life. The dampers can be either automatic or manually operated. If manually operated, they must be equipped with a means to prevent operation of either unit unless the damper is in the full heat or cool position. Purchase them locally.

Ductwork Sizing

The existing or new ductwork must be sized to handle the correct amount of air flow for either heating only or heating and cooling.

Refer to the manufacturer's specification sheet for the equipment in question for the proper air flow characteristics.

Ductwork Insulation

Ductwork that is installed in attic or other areas that are exposed to the outside temperature should be insulated with a minimum of 2 in. of insulation and have an indoor type vapor barrier. Ductwork in other indoor, unconditioned areas should have a minimum of 1 in. insulation with an indoor-type vapor barrier.

Ductwork Connections

The return air can enter through either or both sides or through the bottom of most makes of furnaces.

TABLE 1.5 REMOTE FILTER SIZES (COURTESY OF HEIL-QUAKER CORPORATION)

	Recommended filter sizes Minimum square inches/nominal size filter			
	Disposable type filter low velocity/300 FPM		Cleanable type filter high velocity/500 FPM	
CFM airflow	Minimum surface area (sq. in.)	Recommended nominal size	Minimum surface area (sq. in.)	Recommended nominal size
800	384	20 × 25	231	14 × 20
900	432	20 × 25	260	15 × 20
1000	480	20 × 30	288	14 × 25
1100	528	20 × 30	317	15 × 25
1200	576	14 × 25 (2)	346	16 × 25
1300	624	14 × 25 (2)	375	20 × 25
1400	672	16 × 25 (2)	404	20 × 25
1500	720	16 × 25 (2)	432	20 × 25
1600	768	20 × 25 (2)	461	20 × 25
1700	816	20 × 25 (2)	490	20 × 30
1800	864	20 × 25 (2)	519	20 × 30
1900	912	20 × 30 (2)	548	24 × 25
2000	960	20 × 30 (2)	576	24 × 25

(2) Two Required

By using an optional return air cabinet, the back side may also be used on some makes.

For side connections, cut out an area that is large enough to handle the filter size without making too large a hole in the casing. The procedure is generally outlined in the manufacturer's data.

Filter

The size and type of filter supplied with the furnace will handle the air flow if central air conditioning is used with the furnace.

If external filter grills are used, filters that comply with the specifications in Table 1.5 should be used. The filter size and type must be adequate to handle the CFM requirements based on heating only or heating/cooling applications. See the manufacturer's instructions for the CFM data.

Caution: If the filters provided are suitable for heating applications only, be sure to advise the homeowner so that he or she is aware that the filter size will have to be increased if air conditioning is added.

2

Carrier (58)
High-Efficiency
Gas Furnaces

INTRODUCTION

The desire to conserve energy has created greater use of insulation, improved vapor barriers, weatherstripping, and other energy conserving measures. As a result, buildings are now tighter, resulting in less natural air infiltration and inefficient furnace operation. The condition is further affected by the growing use of kitchen and bathroom exhausts and even fireplaces. Field studies indicate that combustion air starvation, particularly in closet installations, points to a need for positive furnace air supply, plus new guidelines for today's furnace applications.

These instructions cover the minimum combustion air requirements and venting practices. They also reflect current conditions found in the field, and conform to existing national standards and safety codes. In some instances, these instructions exceed certain local codes and ordinances, especially those that have not kept pace with the changing construction practices. Carrier requires these standard procedures as a minimum for safe installation.

The Carrier 58 series furnace is an induced draft furnace. That is, the combustion air is drawn through the combustion zone and the heat exchanger and is exhausted through a blower and vent system (see Figure 2.1).

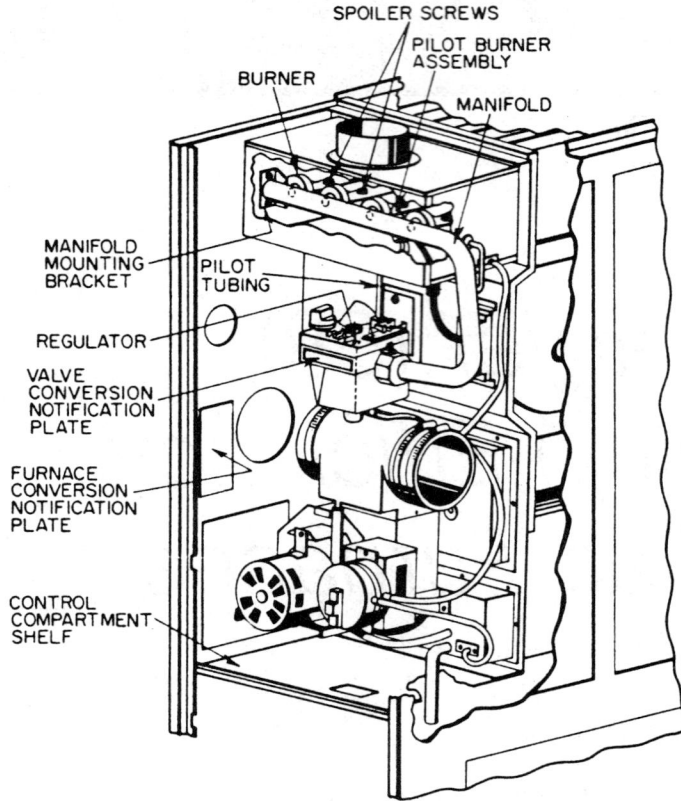

Figure 2.1 Furnace components. (Courtesy of Carrier Corporation.)

Caution: Do not block any openings in the front of the furnace or on the furnace top alongside the vent pipe. These openings provide air for combustion and ventilation. Never store anything on or in this unit. The pipes must terminate through either the roof or a sidewall; roof termination is preferable. Locate the sidewall terminations to prevent damage to shrubs or siding materials.

Table 2.1 gives clearance requirements.

When a previously common-vented system (furnace and water heater) is converted to water heater use only, the vent system may be drastically oversized for the water heater alone. Consult the National Fuel Gas Code for the proper sizing and revise the vent system if necessary.

TABLE 2.1 COMBUSTION-AIR AND VENT TERMINAL CLEARANCES
(COURTESY OF CARRIER CORPORATION)

Location	Clearance
Dryer vent	3 ft
From plumbing vent stack	3 ft
From any opening where vent gases could enter building	12 in.
Above grade and anticipated snow depth	12 in.
Above grade when adjacent to public walkway	7 ft

Warning: Do not install the unit so that indoor air is used for combustion.

Caution: The combustion air must not be taken from inside the structure because that air frequently is contaminated by halogens, which include fluorides, chlorides, bromides, and iodides. These elements are found in aerosols, detergents, bleaches, cleaning solvents, salts, air fresheners, and other household products. Vapors from these products are highly corrosive to gas-fired furnaces, even in extremely low concentrations (as low as 0.5 ppm).

Maintain a minimum of 36 in. between the combustion-air inlet and clothes-drier vent.

Locate the combustion air as far as possible from a swimming pool and swimming pool pump house.

Caution: When the vent pipe is exposed to temperatures below freezing, that is, when it passes through an unheated space or when a chimney is used as a raceway, the pipe must be insulated with 1/2-in. thick Armaflex insulation.

Caution: When the combustion-air pipe is installed above a suspended ceiling, the pipe must be insulated with 1/2-in. Armaflex insulation.

The combustion-air pipe should also be insulated in warm, humid spaces such as basements.

Combustion-Air and Vent Piping

The schedule-40 PVC pipe and fittings must conform to American National Standards Institute (ANSI) standards and American Society for Testing and Materials (ASTM) standards D1785 (schedule-40 PVC), D2661 (ABS-DWV), or D2665 (PVC-DWV). Pipe cement and primer must conform to ASTM D2235 (ABS) or D2564 (PVC). See Table 2.2 for pipe sizing and Figures 2.2–2.4 for exterior piping arrangements.

TABLE 2.2 PIPE DIAMETER (IN.) (COURTESY OF CARRIER CORPORATION)

Pipe length (ft)	Number of 90° elbows (see notes)				
	1	2	3	4	5
5	2	2	2	2	2
10	2	2	2	2	2
15	2	2	2	2	2
20	2	2	2	2	2
25	2	2	2	2	2
30	2	2	2	2	2
35	2	2	2	2	2

NOTES:
1. Assume two 45° elbows equal one 90° elbow.
2. Pipe lengths are "up to and including" the tabulated values.
3. Diameters listed are for schedule-40 PVC, PVC-DWV or ABS-DWV pipe.
4. Long radius elbows are desirable.
5. Elbows and pipe sections supplied in vent terminal kit should not be included in count.

Warning: Solvent cements are combustible. Keep away from heat, sparks, and open flame. Use only in well ventilated areas. DO NOT breathe the vapors. Avoid contact with the skin or eyes.

Warning: All combustion-air and vent pipes must be airtight and water-tight. The pipes must terminate exactly as shown in Figures 2.2–2.4.

The furnace is shipped from the factory assembled for a right-hand vent pipe connection. When a left-hand vent connection is desired, remove the cap

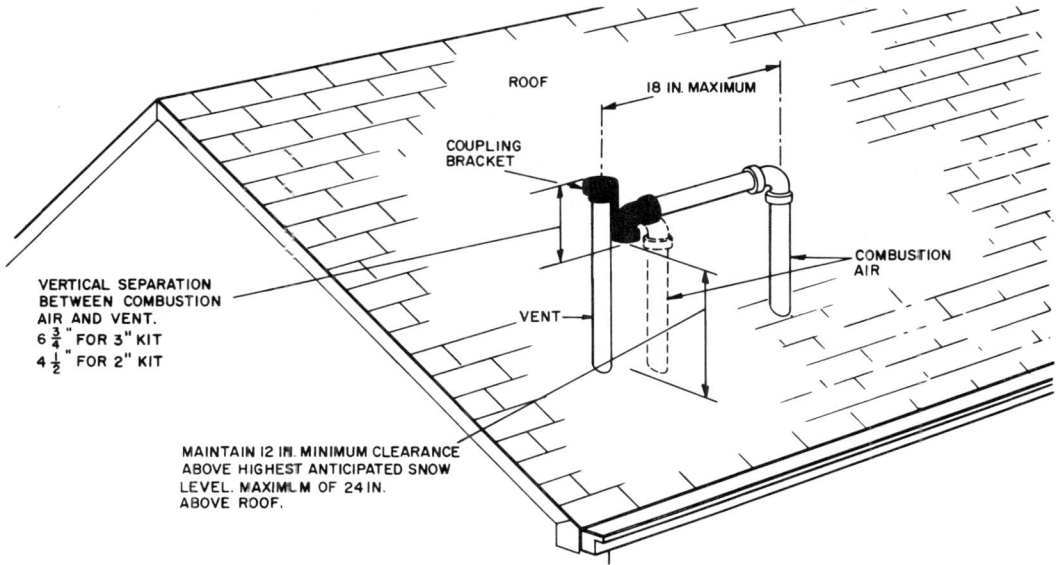

ROOF

18 IN. MAXIMUM

COUPLING
BRACKET

COMBUSTION
AIR

VERTICAL SEPARATION
BETWEEN COMBUSTION
AIR AND VENT.
$6\frac{3}{4}$" FOR 3" KIT
$4\frac{1}{2}$" FOR 2" KIT

VENT

MAINTAIN 12 IN. MINIMUM CLEARANCE
ABOVE HIGHEST ANTICIPATED SNOW
LEVEL. MAXIMUM OF 24 IN.
ABOVE ROOF.

Figure 2.2 Rooftop vent terminal installation. (Courtesy of Carrier Corporation.)

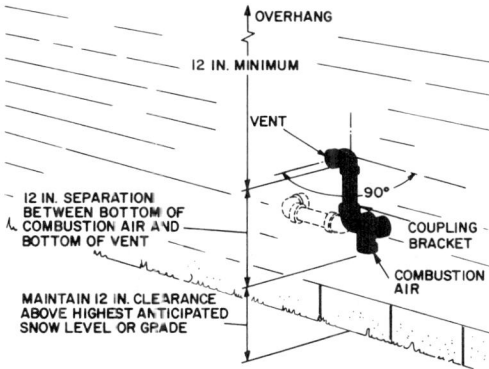

OVERHANG

12 IN. MINIMUM

VENT

12 IN. SEPARATION
BETWEEN BOTTOM OF
COMBUSTION AIR AND
BOTTOM OF VENT

90°

COUPLING
BRACKET

COMBUSTION
AIR

MAINTAIN 12 IN. CLEARANCE
ABOVE HIGHEST ANTICIPATED
SNOW LEVEL OR GRADE

Figure 2.3 Sidewall vent terminal installation (12 in. above snow level or grade). (Courtesy of Carrier Corporation.)

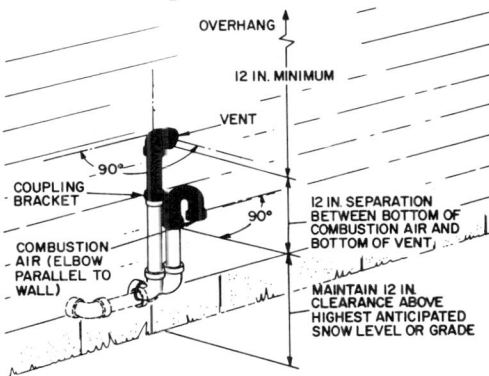

OVERHANG

12 IN. MINIMUM

VENT

90°

COUPLING
BRACKET

90°

COMBUSTION
AIR (ELBOW
PARALLEL TO
WALL)

12 IN. SEPARATION
BETWEEN BOTTOM OF
COMBUSTION AIR AND
BOTTOM OF VENT

MAINTAIN 12 IN.
CLEARANCE ABOVE
HIGHEST ANTICIPATED
SNOW LEVEL OR GRADE

Figure 2.4 Sidewall vent terminal installation (less than 12 in. above snow level or grade. (Courtesy of Carrier Corporation.)

from the left side of the inducer box and install it over the hole in the right side of the box.

Remove the plastic plug from the left-hand casing side panel and install it in the unused hole in the right-hand casing side panel.

Install the tuning valve in the vertical riser of the vent pipe 18 in. above the elbow, as shown in Figure 2.5.

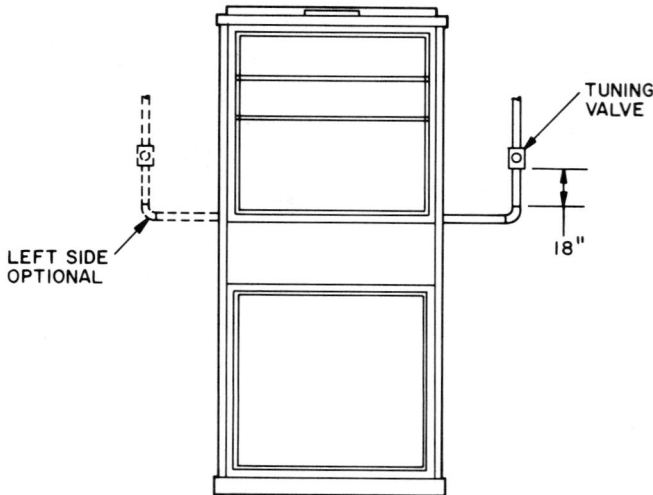

Figure 2.5 Tuning valve location. (Courtesy of Carrier Corporation.)

Use the following steps to install the piping:

1. Attach the factory-supplied flexible coupling to the furnace combustion-air inlet connection and secure it with a stainless-steel hose clamp. Ensure that the factory-supplied, perforated metal combustion-air disk is installed in the flexible coupling.
2. Working from inside the furnace to the outside, cut the PVC pipe to the required length(s).
3. Deburr both the inside and the outside of the pipe.
4. Chamfer the outside edge of the pipe for better distribution of the primer and the cement.
5. Clean and dry all surfaces to be joined.
6. Check the dry fit of the pipe and mark the insertion depth on the pipe.

> **Note:** All pipe should be cut, prepared, and preassembled before any joint is permanently cemented.

7. After all the pipes have been cut and preassembled, apply a generous layer of PVC primer to the pipe-fitting socket and the end of the pipe to

the insertion mark. Quickly apply the PVC cement (over the primer) to the end of the pipe and fitting socket. Apply the cement in a light, uniform coat on the inside of the socket to prevent a buildup of excess cement. Apply a second coat of cement to the end of the pipe.

8. While the PVC cement is still wet, insert the pipe into the socket with a 1/4-turn twist. Be sure that the pipe is fully inserted into the fitting socket.

9. Wipe any excess cement from the joint. A continuous bead of cement is visible around the perimeter of a properly made joint.

10. Handle the pipe joints carefully until the cement sets.

11. Support the piping every 5 ft (minimum) using perforated metal hanging strap. Slope the combustion-air and vent pipes toward the furnace a minimum of 1/4 in. per linear foot with no sags between the hangers.

12. Use appropriate methods to seal openings where the vent and combustion-air pipes pass through the roof or sidewall.

Vent Terminal Kit Installation

The combustion-air and vent pipes must terminate outside the structure. The Carrier accessory vent termination kit for 2-in.-diameter pipe (required) must be installed as shown in Figure 2.2, 2.3, or 2.4. The roof termination, shown in Figure 2.2, is preferred. The kit contains extra parts for the various applications.

Note: The shaded portions of Figures 2.2–2.4 are considered to be a part of the vent terminal and are provided in the vent terminal kit. They should be counted, therefore, in pipe diameter calculations.

Rooftop vent terminal installation (Figure 2.2).

Note: The solid-line arrangement in the figure is the preferred arrangement; the dashed-line arrangement is an option for the combustion-air pipe.

1. Remove one 90° elbow from the elbow and bracket assembly provided in the kit. Loosen the screw so that the other elbow can turn.

2. Loosely install the elbow with the bracket on the combustion-air pipe.

3. Loosely install the pipe coupling, provided in the kit, on a properly cut piece of vent pipe. Position the coupling so that the bracket will mount as shown in the figure.

4. Disassemble the loose pipe fittings. Clean them and apply cement using the procedures described under Combustion-Air and Vent Piping.

5. Install the bracket as shown in the figure.

6. For applications using the combustion-air pipe option (indicated by dashed lines in the figure), install the 180° U-fitting on the end of the pipe instead of the 90° elbow.

Sidewall vent terminal installation.   Install the kit as follows when the combustion-air and vent pipes exit through the sidewall 12 in. or more above the highest anticipated snow level or grade (see Figure 2.3).

> Note: The solid-line arrangement in Figure 2.3 is preferred. The dashed-line arrangement is an option for the vent pipe only.

1. Loosely install the elbow and bracket assembly on the combustion-air and vent pipes.
2. Remove and disassemble the elbow and bracket assembly.
3. Install the elbows as shown in the figure. Clean them and apply the cement using the procedures described under Combustion-Air and Vent Piping.
4. Install the bracket as shown in the figure.
5. Position the vent pipe assembly, maintaining a 12-in. separation. Cement the pipe in the elbow as shown in the figure.
6. For applications using the vent-pipe option (indicated by dashed lines in Figure 2.3), rotate the elbow 90° from the position shown.

Install the kit as follows when the combustion-air and vent pipes exit through a sidewall less than 12 in. above the highest anticipated snow level or grade (see Figure 2.4):

> Note: The solid-line arrangement in Figure 2.4 is preferred. The dashed-line arrangement shown is an option for the vent pipe only.

1. Disassemble the elbow and bracket assembly.
2. Loosely install the bracket on the 180° U-fitting.
3. Loosely install the coupling on the end of the vent pipe.
4. Loosely install the U-fitting and bracket as shown in Figure 2.4. Position the U-fitting so that the open end is against the structure wall.
5. Loosely install the vent-pipe assembly in the coupling as shown in the figure.
6. Check the required dimensions as shown in Figure 2.4.
7. Disassemble the loose pipe fittings. Clean them and apply the cement using the procedures described under Combustion-Air and Vent Piping.
8. Install the bracket as shown in the figure.

Condensate Drain

Route the condensate drain to a nearby floor drain or condensate pump. If a condensate pump is required, it should have a corrosion-resistant impeller and tank.

The 1/2-in.-diameter schedule-40 PVC and CPVC condensate drain piping and fittings must conform to ANSI standards and ASTM D2665 and D22946. The schedule-40 PVC and CPVC cement and primer must conform to ASTM F493 and D2564.

For proper condensate drainage, the furnace must be within 1/2 in. of level. The highest corner of the furnace must not be more than 1/2 in. above the lowest corner.

> **Note:** The furnace contains an internal condensate trap; DO NOT install an external trap.

1. Determine the side of the furnace from which the drain will exit. Cut and preassemble the drain piping (field supplied) directly to an open drain. Refer to Combustion-Air and Vent Piping for instructions on preparing and cementing plastic pipe.
2. Cement the elbow on the pipe assembly (factory supplied) to the condensate trap mounted on the blower housing.
3. Using a second wrench to hold the assembly, attach the first section of the field-supplied drain pipe to the compression coupling provided.
4. When using schedule-40 PVC drain pipe, connect the adapter provided to the end of the pipe installed in the compression fitting.
5. Attach the field-supplied schedule-40 PVC adapter to the threaded factory adapter.
6. Cement the remaining pipe joints.

ELECTRICAL CONNECTIONS

The following is a description of the necessary electrical connections to the furnace:

Line Voltage

> **Important:** Before the electrical connections are made, ensure that the voltage, frequency, and phase correspond to those values specified on the unit rating plate. Also check to be sure that the service provided by your local utility meets the additional load imposed by this equipment. Refer to the unit rating plate for the equipment electrical requirements.

> **Caution:** Do not connect aluminum wire between the disconnect switch and the furnace. Use copper wire only.

Figure 2.6 shows the proper field wiring for both high and low voltage. Make all electrical connections in accordance with the National Electrical Code and any local codes or ordinances that apply.

Use a separate, fused branch electrical circuit containing a properly sized fuse for HACR-type circuit breaker. Locate a disconnecting means in a position in sight and readily accessible from the furnace. The blower door switch may be an acceptable disconnecting means in some areas.

Moving the auxiliary junction box. To move the auxiliary junction box to the right side of the furnace when a right-hand power supply is desired:

1. Remove the 2 screws holding the auxiliary junction box.
2. Drill 2 holes in the same position on the opposite side of the furnace and mount the auxiliary junction box. Rotate the junction box 180° to position the switch in the notch in the casing flange.
3. Plug or cap the unused electrical entry holes in the left side of the casing.

> **Warning:** The furnace must be electrically grounded in accordance with local codes and/or the National Electrical Code ANSI/NFPA 70-1984. DO NOT use gas piping as an electrical ground, to avoid the possibility of personal injury or death.

If the line-voltage wiring to the unit is encased in a nonmetallic sheath, connect the incoming ground wire to the grounding wire inside the furnace junction box. When properly grounded metallic conduit is used, it may serve as the furnace ground.

Low-Voltage Wiring

Make the low-voltage connections at the low-voltage terminal strip according to Figure 2.6.

> **Note:** Use AWG No. 18 color-coded copper thermostat wire for lengths up to 100 ft. Over 100 ft, use AWG No. 16 wire.

> **Important:** Set the thermostat anticipation to match the amperage draw of the gas valve and any other electrical components in the R-W circuit.

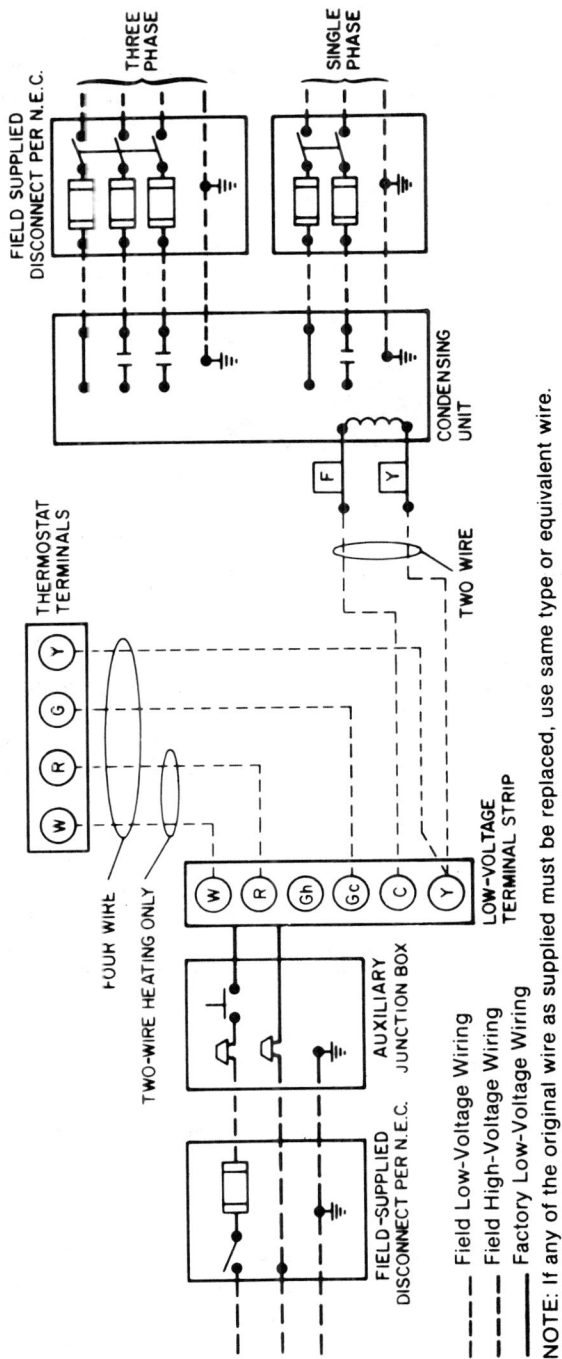

Figure 2.6 Heating and cooling application wiring diagram. (Courtesy of Carrier Corporation.)

THREE PHASE

SINGLE PHASE

FIELD SUPPLIED DISCONNECT PER N.E.C.

CONDENSING UNIT

F

Y

TWO WIRE

THERMOSTAT TERMINALS

Y

G

R

W

FOUR WIRE

TWO-WIRE HEATING ONLY

W

R

Gh

Gc

C

Y

LOW-VOLTAGE TERMINAL STRIP

AUXILIARY JUNCTION BOX

FIELD-SUPPLIED DISCONNECT PER N.E.C.

― ― ― Field Low-Voltage Wiring

― ― ― Field High-Voltage Wiring

――― Factory Low-Voltage Wiring

NOTE: If any of the original wire as supplied must be replaced, use same type or equivalent wire.

Obtain accurate amperage draw readings at the thermostat subbase terminals R and W. Figure 2.7 illustrates an easy method of obtaining actual amperage draw. Take the amperage reading after the blower has started.

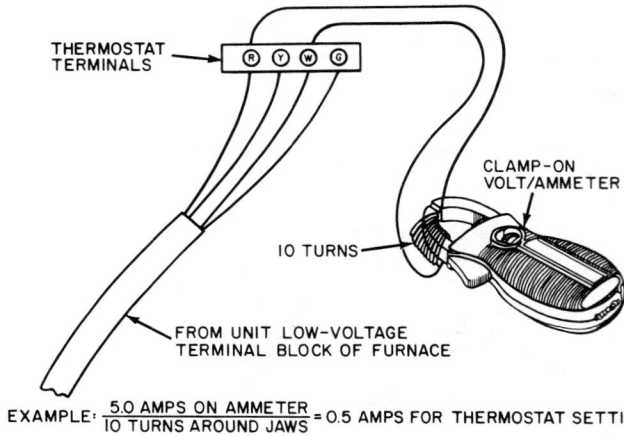

Figure 2.7 Amperage draw check with ammeter. (Courtesy of Carrier Corporation.)

THERMOSTAT
TERMINALS

CLAMP-ON
VOLT/AMMETER

10 TURNS

FROM UNIT LOW-VOLTAGE
TERMINAL BLOCK OF FURNACE

EXAMPLE: $\dfrac{5.0 \text{ AMPS ON AMMETER}}{10 \text{ TURNS AROUND JAWS}} = 0.5$ AMPS FOR THERMOSTAT SETTIN

Locate the room thermostat in the natural circulation path of room air. Avoid locations where the thermostat would be exposed to cold-air infiltration, to drafts from windows, doors, or other openings leading to the outside, or to air currents from warm- or cold-air registers, as well as other locations where the natural circulation of air is restricted, such as behind doors or above or below shelves or mantels.

Do not expose the thermostat to heat from fireplaces, radios, televisions, lamps, or the rays of the sun. Do not mount it on a wall containing warm-air ducts, or flue or vent pipes. Do not choose a location that is inadequately sealed from the attic, crawl space, or basement.

Suitably seal any hole in the plaster or panel through which the wires from the thermostat pass, to prevent drafts from affecting the thermostat.

Blower Control Center

Each furnace has a blower control center to aid the installer or service technician. The low-voltage terminal board is marked for easy connection of the field wiring (see Figure 2.8).

The main furnace control box includes an adjustable blower timing device. The OFF delay can be varied over a range of 80 to 240 seconds by turning the OFF timing adjustment control in the direction indicated on the label attached to the side of the control box. After a change in adjustment, the time-delay circuit must be energized for at least 4 minutes so that the OFF time delay will be the same as during normal furnace operation. The OFF timing

BLOWER OFF-TIME ADJUSTMENT CONTROL

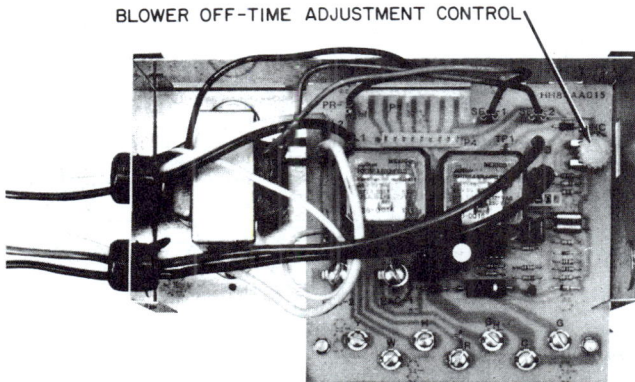

Figure 2.8 Blower control center.
(Courtesy of Carrier Corporation.)

adjustment is factory set for a delay of approximately 240 seconds (see Figure 2.8). The ON time delay is not adjustable; it is set for 60 seconds.

SEQUENCE OF OPERATION

The following is a description of both the heating and the cooling cycles:

Heating Cycle

Refer to Figure 2.9.

1. When the blower door is in place, 115-V current is supplied through the blower door interlock switch 9G. The transformer 1A is energized, supplying 24 V to the heating blower relay 2E, which opens the normally closed blower relay contacts 2E in the low-speed circuit of the blower motor 3D.
2. When the wall thermostat calls for heating, the R and W circuit closes, supplying electrical power to the 24-V safety circuit containing the limit switch 7H1.
3. The inducer-motor relay coil 2D is energized and relay contacts 2D in the 115-V circuit close, starting the inducer motor 3A. Simultaneously, another set of contacts in the inducer-motor relay 2D closes in the 24-V circuit, locking in the inducer-motor relay coil 2D. The coil is locked in until the R and W circuit or the safety circuit opens.
4. As the inducer motor 3A comes up to speed, the flow-sensing switch 7V energizes the HOLD coil of the gas valve 5F. Simultaneously, the PICK coil of the gas valve 5F and the time delay circuit of the spark generator 6J are energized through the fusible link 11C.

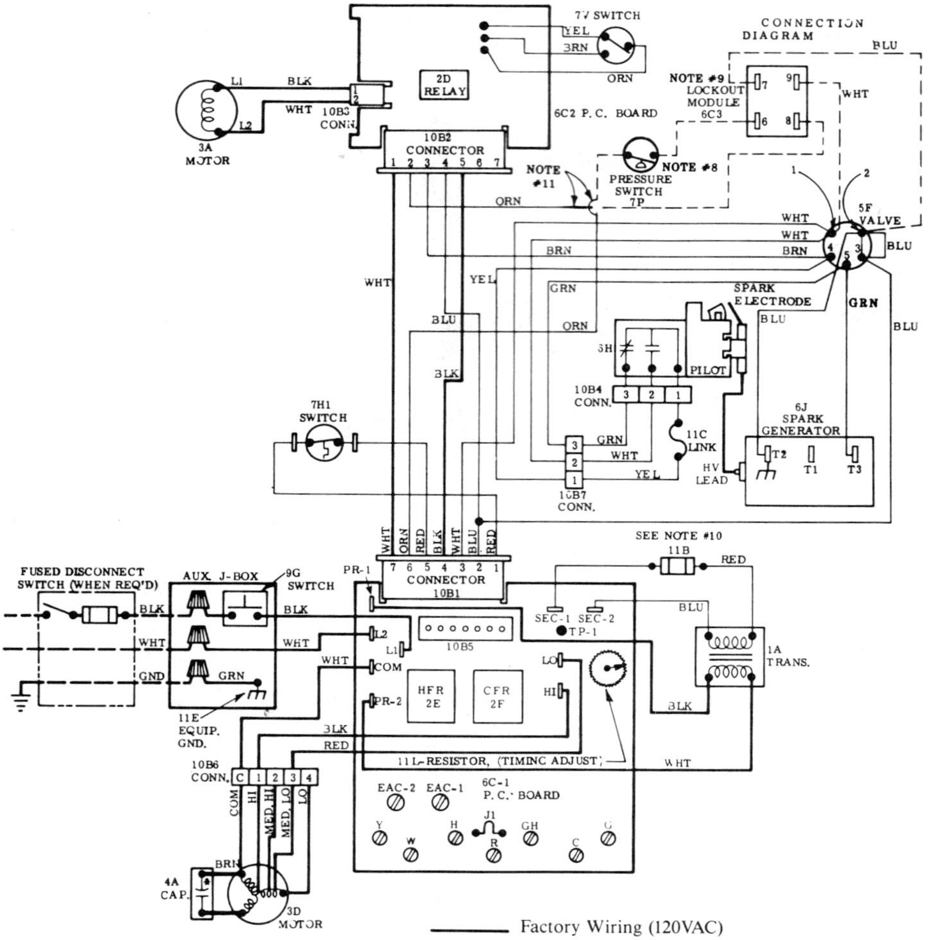

Figure 2.9 Schematic wiring diagram. (Courtesy of Carrier Corporation.)

Factory Wiring (120VAC)
Factory Wiring (24VAC)
Field Wiring (120VAC)
Conductors on 6C1 (Furnace Control Board)
Conductors on 6C2 (Inducer Control Board)
Screw Terminal for Field Wiring
1/4 Inch Quick Connect Terminals

Legend:
1A–Transformer, 120VAC, 24VAC
2D–Relay, Inducer Motor DPST (N.O.)
2E–Relay, Heating Blower (HFR) DPST (N.C.)
2F–Relay, Cooling Blower (CFR) DPDT
3A–Motor, Inducer
3D–Motor, Blower
4A–Capacitor, Run
5F–Valve, Gas (Redundant)
6C1–Board, P.C. (Furnace Control)
6C2–Board, P.C. (Inducer Control)
6C3–Lockout Module
6H–Switch, Pilot Flame Sensing SPDT (740A)
6J–Generator, Spark (Solid State)
7H1–Switch, Limit SPST (N.C.)

7P–Pressure Switch, N.O.
7V–Switch, Flow Sensing SPDT
9G–Switch, Blower Door Interlock SPST (N.O.)
10B1–Connector, Edge (Furnace Control Board)
10B2–Connector, Edge (Inducer Control Board – 7 CKT.)
10B3–Connector, Edge (Inducer Control Board – 2 CKT.)
10B4–Connector, Pilot
10B5–Connector, Factory Test (Not for Field Use)
10B6–Connector, Blower Motor
10B7–Connector, Pilot Splice
11B–Fuse, in Line 2Amp (When Used)
11C–Link, Fusible (Over Temperature)
11E–Ground, Equipment
11L–Resistor, Adjustable (Off Time)
TP1–Test Point

44

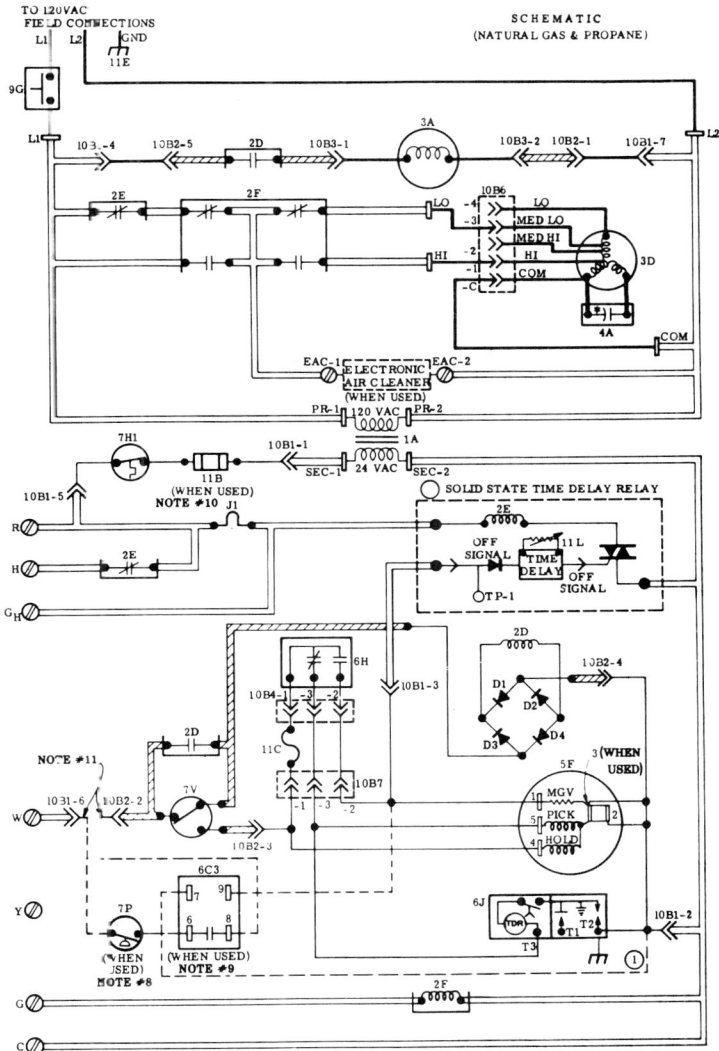

SCHEMATIC
(NATURAL GAS & PROPANE)

TO 120VAC FIELD CONNECTIONS

L1 L2 GND
11E

9G

L1 103.-4 10B2-5 2D 10B3-1 3A 10B3-2 10B2-1 10B1-7 L2

2E 2F LO 10B6
LO
MED LO
MED HI
HI HI 3D
COM
4A COM

EAC-1 ELECTRONIC EAC-2
AIR CLEANER
(WHEN USED)
PR-1 120 VAC PR-2

7H1 10B1-1 1A
SEC-1 24 VAC SEC-2
11B
(WHEN USED)
NOTE #10 J1

10B1-5 SOLID STATE TIME DELAY RELAY
2E
R OFF
SIGNAL TIME 11L
2E DELAY OFF
H SIGNAL
TP-1
G

6H 2D 10B2-4
10B4 D1
D2
11C D3 D4
10B7 3 (WHEN USED)
5F
NOTE #11 MGV
2D PICK
W HOLD
1031-6 10B2-2
7V
10B2-3

6C3
7P 9 6J
Y 6 8 TDR T2 T1 10B1-2
(WHEN (WHEN USED) T3
USED) NOTE #9
NOTE #8 1

2F
G
C

Notes:
1. T2 Internally Connected to Equipment Ground Spark Generator (6J) Mounting Screws.
2. Relay (2E) Contacts are Normally Closed Until 120VAC is Applied to Furnace.
3. If Any of the Original Wire As Supplied Must Be Replaced, it Must Be Replaced with AWM (105°C) Wire or its Equivalent.
4. Blower Motor (3D) and Inducer Motor (3A) Have a Thermal Overload Switch.
5. Blower Motor (3D) Factory Speed Selections Are for Average Conditions. See Installation Instructions for Details on Optimum Speed Selection.
6. Use Copper Wire Only Between the Disconnect Switch and the Unit.
7. Symbols are an Electrical Representation Only.
8. Pressure Switch (7P) is Used for Propane Only.
9. Lockout Module (6C3) is Used for Propane or 100% Shutoff Natural Gas.
10. 2Amp Fuse in Secondary of Transformer for Canadian Units Only.
11. Factory Connected When Accessory Not Used.

45

5. When the PICK coil of the gas valve 5F is energized, gas flows to the pilot. The normally open time-delay circuit closes after a 10-second purge delay, energizing the spark generator 6J. The pilot is ignited by the spark generator.

6. The PICK coil of the gas valve and the spark generator 6J are deenergized when the normally closed pilot-sensing switch contacts 6H open. The normally open flame-sensing switch contacts 6H close 5 to 20 seconds later, energizing the MGV (main operator) of the gas valve 5F. The gas valve 5F opens approximately 10 seconds later, allowing gas to flow to the main burners. The pilot flame immediately ignites the gas.

7. Simultaneously, the time-delay circuit 11L in the blower control center is energized. Approximately 50 seconds after the gas valve 5F opens, the heating relay coil 2E is deenergized, which closes the 115-V contacts of the heating relay 2E, starting the blower motor 3D on the heating speed. The H terminal is energized with 24-VAC when the blower motor operates on the heating speed. The electronic air cleaner (EAC) terminals energize with line voltage when the blower operates on either the heating or the cooling speed.

8. When the thermostat is satisfied, the circuit between R and W is broken, deenergizing the gas valve 5F, the inducer-motor relay 2D, and the solid-state time-delay circuit on the main printed circuit board. Gas flow to the pilot and the main burner stops immediately. After approximately 80 to 240 seconds, depending on the blower OFF time adjustment, the relay 2E is energized and the blower 3D stops.

> **Note**: After a brief interruption of either the gas or the electric supply, the furnace will not resume operation until the contacts of the pilot-flame sensing switch 6H move from the normally open to the normally closed position.

Cooling Cycle

Refer to Figure 2.9.

1. The wall thermostat calls for cooling.

2. The R, G, and Y circuits are energized. Simultaneously, the R and Y circuit starts the outdoor condensing unit, and the R and G circuit energizes the cooling relay coil 2F. This closes the normally open contacts 2F, energizing the cooling speed of the blower motor 3D and opening the normally closed contacts of the cooling relay 2F. The electronic air cleaner (EAC) terminals are energized with line voltage whenever the blower is operating on either the heating or the cooling speed.

FILTER

The following is a description of the different procedures for installing filters:

Filter Arrangement

The filter is supplied in the furnace blower compartment. For bottom inlet application, the filter should be installed as shown in Figure 2.10.

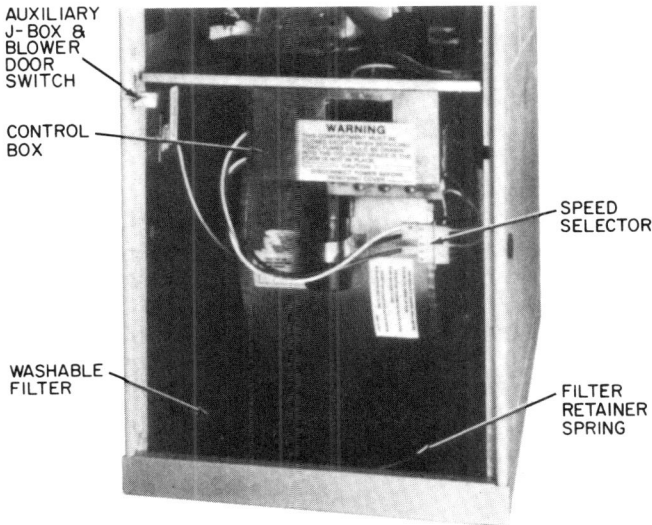

Figure 2.10 Filter installation for bottom inlet. (Courtesy of Carrier Corporation.)

Note: Remove and discard the bottom closure panel when the bottom inlet is used.

Figure 2.11 shows the opening sizes for a side inlet application. Remove the filter and retainer spring from the bottom opening. Install the retainer spring in the holes provided: one in the blower shelf and the other in the retainer spring bracket (see Figure 2.12). Move the unused retainer bracket to the opposite side, near the front of the furnace, to support the filter.

Bottom Closure Panel

Refer to Figure 2.13. When the side inlet is used, the bottom opening must be properly sealed by installing the bottom closure panel shipped under the filter in each furnace. To install the bottom closure panel, perform the following steps:

FURNACE SIZE	040	060	080	100	120
DIMENSIONS (in.)					
Width A	17½	17½	17½	21	24½
Combustion Air Vent Diameter B	1½ or 2	1½ or 2	2 or 2½	2, 2½ or 3	3
Outlet Duct Opening (C x 19)	15⅞	15⅞	15⅞	19⅜	22⅞
Bottom Return Opening (D x 26)	15	15	15	18½	22
Filter Size, 27¾ x 1 x	15⅞	15⅞	15⅞	19⅜	22⅞

Left diagram labels:

1½" DIA GAS CONN. (L.H.)
3⅛" DIA VENT CONN.
⅞" DIA ELEC CONN.
⅜" DIA THERMOSTAT CONN.
⅞" DIA CONDENSATE DRAIN
SIDE INLET

16⅛", 11⅛", 25 13/16", 23½", 26 9/16", 27 5/16"

Right diagram labels:

ALT GAS CONN.
VENT CONN. (R.H.)
ALT ELEC CONN.
ALT THERMOSTAT CONN.
ALT CONDENSATE DRAIN
SIDE INLET

46 3/16", 16⅛", 14½" TYP, 19", C, B, 26 5/16", 23½", 26 9/16", 27 5/16", 23¾" TYP, 28½", D, A

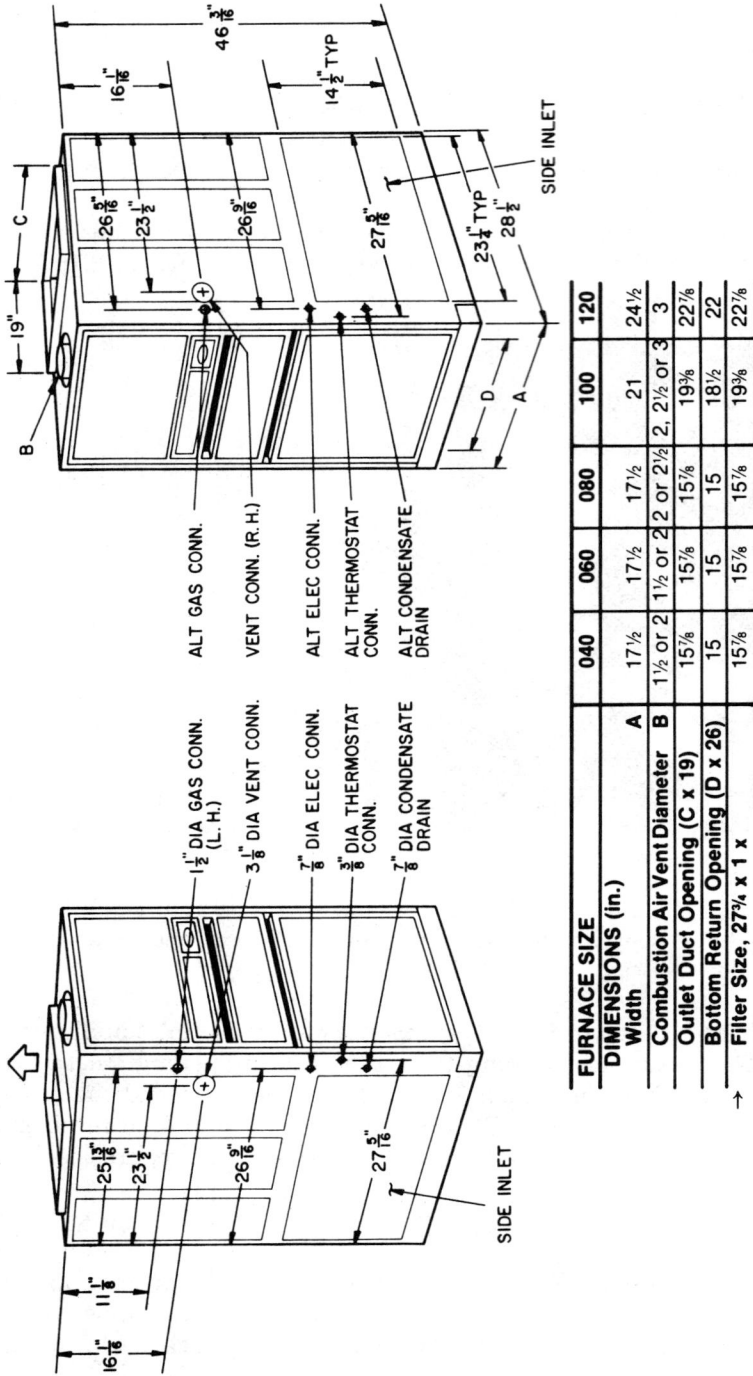

Figure 2.11 Unit dimensions. (Courtesy of Carrier Corporation.)

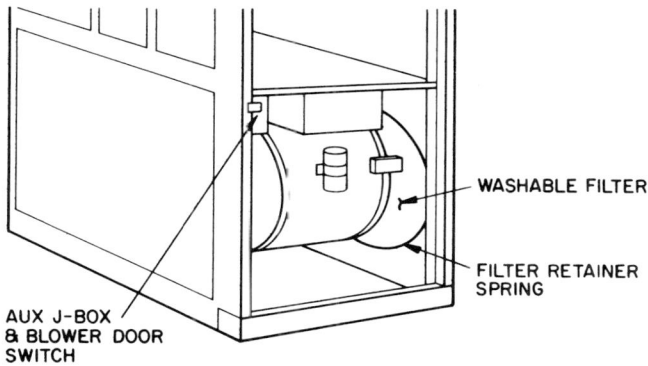

Figure 2.12 Filter installed for side inlet. (Courtesy of Carrier Corporation.)

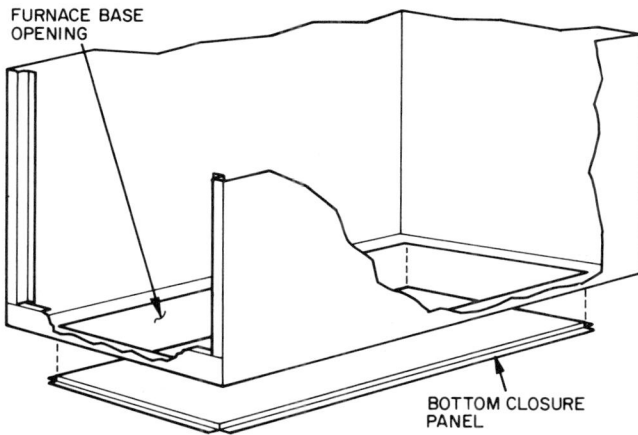

Figure 2.13 Installing bottom closure panel. (Courtesy of Carrier Corporation.)

1. After the filter has been installed for a side return, remove the bottom closure.
2. With the furnace either tilted or raised, install the panel, insulation up, in the opening from the bottom of the furnace.

START-UP, ADJUSTMENT, AND SAFETY CHECK

The following steps are recommended for these procedures:

Adjusting Tuning Valve

Before firing the furnace, adjust the pressure drop through the heat exchanger for maximum efficiency, following the steps below.

> **Caution**: Be sure that the gas supply to the furnace is turned off.

1. Install a field-supplied plastic tee between the pressure tap on the bottom of the gas valve and pressure tube, as shown in Figure 2.14.

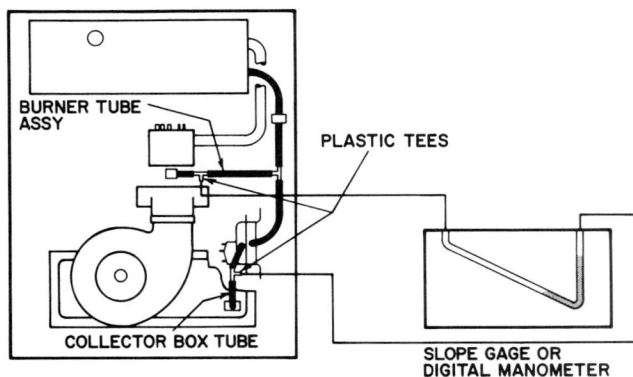

Figure 2.14 Tuning valve adjustment. (Courtesy of Carrier Corporation.)

2. Install a second field-supplied plastic tee between the pressure switch and the pressure tube from the collector box (see Figure 2.14).
3. Connect a slope gauge to the tees (see Figure 2.14).
4. Close the R to W circuit to start the inducer motor.
5. Adjust the tuning valve to obtain 0.83 ± 0.01 in. negative water column.
6. Remove the handle from the tuning valve and recheck the pressure. Store the handle in a safe place.
7. Open the R to W circuit.
8. Disconnect the slope gauge.
9. Remove the plastic tees and reconnect the factory pressure tubes to the gas valve and the pressure switch.
10. Turn on the gas supply to the furnace.

Ignition System Check

When all the connections have been checked, light the furnace using the procedure outlined on the lighting instruction plate attached to the furnace. When lighting the furnace for the first time, however, perform the following additional steps:

1. If the gas supply was not purged before connecting to the furnace, the line will be full of air. Loosen the ground joint union and allow the supply

line to purge until a gas odor is detected. Never purge gas lines into a combustion chamber. Immediately upon detection of the gas odor, retighten the union. After 5 minutes, light the furnace in accordance with the instructions on the furnace rating plate.

2. The main burners should light within 25 to 75 seconds after the pilot. If the main burners do not light within the prescribed time period, adjust the pilot flame, allow the pilot to cool for 5 minutes, and then repeat the time check (see Figure 2.15).

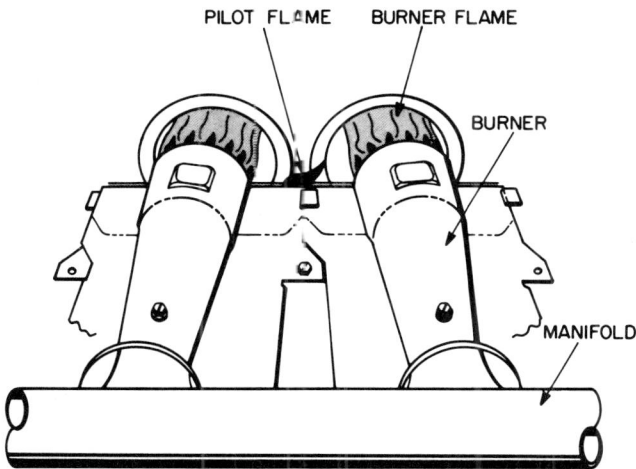

PILOT FLAME   BURNER FLAME

BURNER

MANIFOLD

Figure 2.15 Proper pilot and burner flames. (Courtesy of Carrier Corporation.)

3. Locate the pilot-adjusting screw on top of the valve:
   a. Remove the cap screw; turn the pilot-adjusting screw counterclockwise to decrease the burner-on time delay, or clockwise to increase the burner-on time delay.
   b. Replace the cap screw.

Gas Input

> **Note**: Be sure that the reference pressure tube, combustion-air and vent pipes, and burner enclosure front plate are in place when clocking the gas meter.

1. Determine the gas input:
   a. Turn off all other gas appliances and pilots.
   b. Measure the time (in seconds) for the gas meter test dial to complete one revolution.
   c. Refer to Table 2.3 for the cubic feet of gas per hour.
   d. Multiply the cubic feet per hour by the heating value of the gas (Btu/ft^3) obtained from the local utility.

TABLE 2.3   GAS RATE (FT³/HOUR) (COURTESY OF CARRIER CORPORATION)

Seconds for one revolution	Size of test dial			Seconds for one revolution	Size of test dial		
	1 cu ft	2 cu ft	5 cu ft		1 cu ft	2 cu ft	5 cu ft
10	360	720	1800	50	72	144	360
11	327	655	1636	51	71	141	355
12	300	600	1500	52	69	138	346
13	277	555	1385	53	68	136	340
14	257	514	1286	54	67	133	333
15	240	480	1200	55	65	131	327
16	225	450	1125	56	64	129	321
17	212	424	1059	57	63	126	316
18	200	400	1000	58	62	124	310
19	189	379	947	59	61	122	305
20	180	360	900	60	60	120	300
21	171	343	857	62	58	116	290
22	164	327	818	64	56	112	281
23	157	313	783	66	54	109	273
24	150	300	750	68	53	106	265
25	144	288	720	70	51	103	257
26	138	277	692	72	50	100	250
27	133	267	667	74	48	97	243
28	129	257	643	76	47	95	237
29	124	248	621	78	46	92	231
30	120	240	600	80	45	90	225
31	116	232	581	82	44	88	220
32	113	225	563	84	43	86	214
33	109	218	545	86	42	84	209
34	106	212	529	88	41	82	205
35	103	206	514	90	40	80	200
36	100	200	500	92	39	78	196
37	97	195	486	94	38	76	192
38	95	189	474	96	38	75	188
39	92	185	462	98	37	74	184
40	90	180	450	100	36	72	180
41	88	176	439	102	35	71	178
42	86	172	429	104	35	69	173
43	84	167	419	106	34	68	170
44	82	164	409	108	33	67	167
45	80	160	400	110	33	65	164
46	78	157	391	112	32	64	161
47	76	153	383	116	31	62	155
48	75	150	375	120	30	60	150
49	73	147	367				

Example

Btu heating input = Btu/ft^3 times the ft^3/h
Heating value of gas = 1070 Btu/ft^3
Time for one revolution of a 2-ft^3 dial = 72 seconds
Gas rate = 100 ft^3/h (from Table 2.3)
Btu heating input = 1070 × 100 = 107 000 Btuh

    e. The measured gas input must not exceed the input on the unit rating plate.

2. To adjust the input rate:
    a. Remove the burner enclosure front and cap that conceals the adjustment screw for the gas valve regulator.
    b. Turn the adjusting screw either counterclockwise (out) to decrease the input rate or clockwise (in) to increase the rate. When adjusting the input rate, *do not* change the manifold pressure more than 0.3 in. water column (wc). Make any major adjustments by changing the main burner orifices (see Figure 2.15).

Note: The manifold pressure must always be measured with the burner enclosure front removed. The gas meter must always be clocked with the burner enclosure front installed.

    c. Replace the burner enclosure front and measure the adjusted gas input rate using the method outlined in step 1.
    d. Replace the cap that conceals the gas valve regulator adjustment screw.

Caution: Be sure that the burner enclosure front is in place after all the adjustments have been made.

3. Look through the sight glass in the burner enclosure and check the burner and pilot flame. The main burner flame should be clear blue and almost transparent. The pilot flame should be well defined (see Figure 2.15).
4. High altitudes: The input ratings apply for altitudes up to 2000 ft. Ratings for altitudes above 2000 ft must be 4% less for each 1000 ft above sea level.

Temperature Rise

Do not operate the furnace outside the range of temperature rise specified on the unit rating plate. Determine the air temperature rise as follows:

1. Place thermometers in the return and supply air ducts as near the furnace as possible. The thermometers must not "see" the heating element or radiant heat may affect the readings. This practice is particularly important with straight-run ducts.

2. When the thermometer readings stabilize, subtract the return-air temperature from the supply-air temperature to determine the temperature rise.

3. Adjust the temperature rise by adjusting the blower speed. Increase the blower speed to reduce the temperature rise. Decrease the blower speed to increase the temperature rise.

Thermostat Heat Anticipator Adjustment

The thermostat heat anticipator must be set to match the amperage draw of the components in the R-W circuit. Accurate amperage draw measurements can be obtained from the thermostat subbase terminals R and W. Figure 2.16 illustrates an easy method of obtaining these measurements. The amperage reading should be taken after the blower has started on its heating speed.

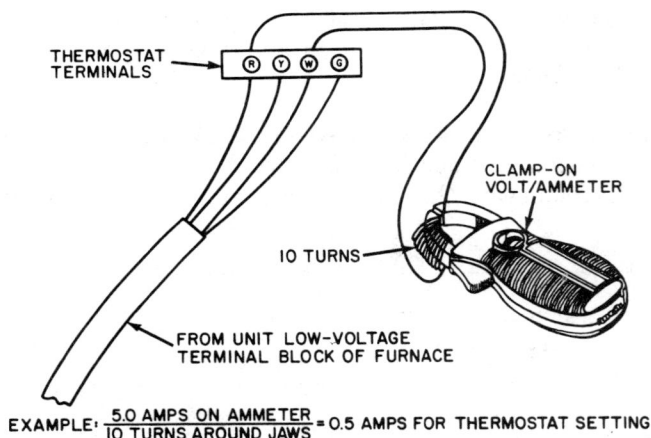

THERMOSTAT TERMINALS

CLAMP-ON VOLT/AMMETER

IO TURNS

FROM UNIT LOW-VOLTAGE TERMINAL BLOCK OF FURNACE

EXAMPLE: $\dfrac{5.0 \text{ AMPS ON AMMETER}}{\text{IO TURNS AROUND JAWS}}$ = 0.5 AMPS FOR THERMOSTAT SETTING

Figure 2.16 Amperage draw measurements. (Courtesy of Carrier Corporation.)

Safety Check of Limit Control

This control shuts off the combustion gas supply and energizes the circulating-air blower motor if the furnace overheats.

The recommended method of checking this limit control is to gradually block off the return air after the furnace has been operating for at least 5 minutes. As soon as the limit control functions, the return-air opening should be unblocked to permit normal air circulation. By using this method to check the limit control, proper functioning can be determined and the furnace will fail-safe if the circulating air supply is restricted or if the motor fails. If the limit control does not function during this test, identify and correct the cause of malfunction.

Safety Check of Flow-Sensing Switch

This control proves operation of the draft inducer. Check the switch operation as follows:

1. Turn off the 115-V power to the furnace.
2. Remove the control access panel and disconnect the inducer motor lead wires from the inducer printed-circuit board.
3. Turn on the 115-V power to the furnace.
4. Close the thermostat switch as if making normal furnace start. When the flow-sensing switch is functioning properly, the gas should not flow to the pilot and the ignitor should not operate. If either the pilot gas flow or the ignitor operation occurs when the inducer motor is disconnected, shut the furnace down immediately. Determine the cause of malfunction and correct the condition.
5. Turn off the 115-V power to the furnace.
6. Reconnect the inducer-motor leads, reinstall the control access panel, and turn on the 115-V power supply.

Adjustment of Blower Speed

> **Warning:** Disconnect and tag the electrical power before changing the speed tap, to avoid possible personal injury.

To change the motor speed taps, remove the motor tap lead. See Figure 2.10 and Table 2.4 and relocate it to the desired terminal on the plug-in terminal block/speed selector.

TABLE 2.4 SPEED SELECTOR (COURTESY OF CARRIER CORPORATION)

Speed	Tap no.	Color
Common	C	White
Hi	1	Black
Med-Hi	2	Yellow
Med-Low	3	Blue
Low	4	Red

*White wire to common, black wire to cooling speed selection red wire to heating speed selection.

> **Caution:** When adjusting the blower speed, check to see that the temperature rise across the heat exchanger does not exceed that specified on the rating plate of the furnace.

Automatic Gas Control Valve

These units are equipped with automatic gas control valves. If it was not checked when lighting the main burner, check for proper operation of this valve by moving the room thermostat pointer above and below the room temperature. The main burners should light when the pointer is above, and should shut off when the pointer is below the room temperature setting.

> **Note:** For Model 646 gas valves: For ease of adjusting the pilot flame, disconnect the wire from terminal 1. This prevents main burner ignition and allows time to adjust the pilot. Be sure to reconnect the power lead after making this adjustment. Check the flame characteristics after replacing the burner closure front.

CARE AND MAINTENANCE

> **Caution:** To prevent both possible damage to the equipment and personal injury, all maintenance should be performed only by qualified personnel.

> **Warning:** Never store any flammable chloride- or halogen-containing compounds near the furnace, to avoid the possibility of fire, personal injury, or death. Do not store:
> 1. Spray or aerosol cans, rags, brooms, dust mops, vacuum cleaners, or other cleaning tools.
> 2. Soap powders, bleaches, waxes, or other cleaning compounds; plastic items or containers; gasoline, kerosene, cigarette lighter fluid, dry-cleaning fluids, or other volatile fluids.
> 3. Paint thinners and other painting compounds.
> 4. Paper bags or other paper products.

For continuing high performance, and to minimize equipment failure, it is essential that periodic maintenance be performed on this equipment. The ability to perform maintenance on this equipment properly requires certain mechanical skills and tools. It is the responsibility of the service technician to obtain these skills and tools.

> **Warning**: Turn off and tag the gas and electrical supplies to the unit before performing any maintenance or service on the unit to avoid the possibility of personal injury. Follow the relighting instructions attached to the furnace.

Minimum Maintenance Requirements

The following steps are the minimum amount of maintenance that should be done for this furnace:

1. Check and clean or replace the air filter each month or as required.
2. Check the blower motor and wheel for cleanliness and lubrication at the beginning of each heating and cooling season. Clean and lubricate as necessary. See Blower Motor and Wheel list below.
3. Check the electrical connections for tightness and the controls for proper operation each heating season. Service as necessary.
4. Check for proper condensate drainage.
5. Check for blockages of the combustion-air and vent pipes.

> **Warning:** As with any mechanical equipment, personal injury can result from contact with sharp metal edges and other objects. Be careful to avoid such contact when removing or replacing parts.

Air Filter

Disconnect and tag the electrical power before removing the access panels. To clean or replace the air filter use the following procedure:

1. Remove the blower access panel.
2. Release the filter retaining spring from behind the flange of the furnace casing (see Figures 2.10 and 2.12).
3. Slide out the filter.
4. Clean the filter with cold tap water. Spray the water in the direction opposite to that of the air flow.
5. Rinse and let dry. *Do not* oil or coat the filter.

6. Place the dry filter in the furnace with the cross-sectional binding facing the blower.

Blower and Wheel

For long life, economy, and high efficiency, clean any accumulated dirt and grease from the blower wheel and motor annually. The following steps should be performed only by qualified service personnel.

Lubricate the motor every 5 years if the motor is used on intermittent operation (thermostat FAN switch in AUTO position) or every 2 years if the motor is in continuous operation (thermostat FAN switch in the ON position).

> **Caution:** Disconnect and tag the electrical supply before removing the access panels.

Clean and lubricate as follows:

1. Remove the access panels.
2. Remove the control box from the bottom side of the blower shelf and position it out of the way.
3. Remove the electrical leads from the numbered side of the blower speed selector (see Figure 2.10 and Table 2.4). Note the location of the wires to aid in the reassembly process.
4. Using a second wrench to hold the line in place, disconnect the drain pipe at the coupling in the blower compartment.
5. Loosen the hose clamps and remove the 7/8-in.-diameter drain hose.
6. Loosen the hose clamp and disconnect the 5/8-in.-diameter drain hose at the bottom of the inducer housing located under the blower shelf.
7. Remove the screws securing the drain trap assembly.
8. Remove the screws securing the blower assembly to the blower shelf and slide the blower assembly out of the furnace.
9. Squeeze the side tabs of the blower speed selector and pull them from the blower housing bracket.
10. Loosen the screw in the strap holding the motor capacitor to the blower housing, and slide the capacitor from the strap.
11. Mark the blower wheel position on the shaft, and the motor support location on the motor before disassembly to ensure proper assembly.
12. Loosen the set screw holding the blower wheel on the motor shaft.
13. Remove the bolts holding the motor mount to the blower housing and

slide the motor and mount from the housing. Disconnect the ground wire that is attached to the blower housing before removing the motor.

14. Lubricate the motor:
    a. Remove the dust caps or plugs from the oil ports that are located at each end of the motor. If the motor does not have these caps or plugs, the bearings are sealed and need no further lubrication.
    b. Use a good grade of SAE 20 nondetergent motor oil. Add one teaspoon (5 ml, 3/16 oz, or 25 drops) to each oil port. Other types or grades of oil could damage the motor. Excessive oiling can cause premature bearing failures.
    c. Allow time for the total quantity of oil to be absorbed by each bearing.
    d. After oiling the motor, wipe any excess oil from the motor housing.
    e. Replace the dust caps or plugs on the oil ports.
15. Remove the blower wheel from the housing.
    a. Mark the blower wheel orientation and cutoff location to ensure proper reassembly.
    b. Remove the screws securing the cutoff plate and remove the cutoff plate from the housing.
    c. Lift the blower wheel from the housing through the opening.
16. Use a vacuum with a soft brush attachment to clean the blower wheel and motor. *Do not* disturb the balance weights (clips) on the blower wheel vanes. *Do not* drop or bend the blower wheel because the balance will be affected.
17. Reassemble the blower by reversing steps 15a–c. The wheel must be repositioned properly, with the motor oilers pointing up, when the motor is installed.
18. Reassemble the motor and blower by reversing steps 9–13. If the motor has a ground wire, it must be reconnected. The blower wheel must be centered in the blower housing. Spin the blower wheel to check the clearance.
19. Reinstall the blower assembly in the furnace.
20. Inspect the drain trap and hoses to ensure that they are not blocked or restricted. Replace the drain trap and hoses. Tighten the hose clamps.
21. Use a second wrench to hold the assembly while attaching the drain pipe and tightening the compression coupling.
22. Connect the electrical leads to the blower speed selector. Note that the connections are polarized for correct assembly. Do not force.
23. Reinstall the control box on the bottom side of the blower shelf.
24. Turn on the electrical power and check for proper rotation of the blower and speed changes between the heating and the cooling cycles. Operate the unit for 5 minutes and check for condensate leaks.

Cleaning Heat Exchangers

If it becomes necessary to clean the heat exchangers, proceed as follows:

1. Turn off and tag the gas and electrical supplies to the furnace.
2. Remove the control and blower access panels.
3. Loosen the hose clamps on the combustion-air pipe and move the pipe aside.
4. Use a second wrench to hold the assembly in place while disconnecting the gas supply at the ground joint union. Remove the gas pipe from the valve.
5. Disconnect the pilot leads at the 3-circuit connector outside of the burner enclosure.
6. Disconnect the high-voltage lead at the spark generator.
7. Disconnect the electrical wires from the gas valve.
8. Disconnect the pressure tubing from the right side of the burner enclosure and from the bottom of the gas valve.
9. Remove the burner enclosure front.
10. Remove the diffuser from inside the top of the burner enclosure. Remove the screws that secure the enclosure to the cell panel. These screws are located inside the burner enclosure.
11. Remove the gas control assembly from the furnace carefully, so that the cell inlet panel gasket is not damaged.
12. Loosen the hose clamps at the vent pipe connection; disconnect the vent pipe and position it out of the way.
13. Disconnect the edge connector from the inducer control box.
14. Disconnect the edge connector from the main control box at the blower shelf.
15. Remove the screws securing the main control box to the blower shelf and position the control box out of the way.
16. Loosen the hose clamp and remove the drain tube from the inducer housing located on the bottom side of the blower shelf.
17. Remove the mounting screws securing the inducer assembly to the collector box; remove the inducer assembly.
18. Remove all old sealant from the parts (when used).
19. Loosen the hose clamp and remove the drain tube from the inducer outlet box.
20. Remove the screws securing the coupling box to the cell panel and remove the box from the furnace. Remove all old sealant from the parts.
21. Remove the choke plate (if used).

22. Loosen the hose clamp and remove the 7/8-in. drain tube from the trap.
23. Place a bucket under the 7/8-in. drain tube in the blower compartment.
24. Using a garden hose, flush each cell of the condensing heat exchanger with water. Use care not to spray water onto the interior surfaces of the control compartment. Dry all surfaces.
25. Using a field-provided small wire brush, steel snake cable, reversible electric drill, and a vacuum cleaner, clean the primary heat exchanger cells.
26. Connect the 7/8-in. drain tube to the trap and tighten the hose clamp.

Reassemble the Furnace

Use the following steps to reassemble the furnace:

1. Install the choke plate (if used).
2. Apply a sealant-releasing agent (Pam) to the flange of the coupling box.
3. Apply a generous bead (3/16-in.-diameter) of GE RTV 162 sealant or Dow-Corning RTV 738 sealant (*no substitute is permissible*) to the flange of the coupling box over the releasing agent. GE RTV 162 and Dow-Corning RTV 738 are available from the factory distributors.
4. Being careful not to smear the sealant, position the coupling box so that the slot in the insulation is on the left-hand side and install the coupling box.
5. Place the small round gasket(s) between the blower shelf and the inducer housing. If the inducer housing uses an inlet gasket, disregard steps 6 and 7.
6. Apply the sealant-releasing agent (Pam) to the collector box.
7. Apply a 1/8-in.-diameter bead of GE RTV 162 or Dow-Corning RTV 738 sealant to the back of the inducer housing. Apply the sealant around the inlet air opening. Apply the sealant about 1/4-in. from the edge of the inlet air opening.
8. Using the stainless steel screws, install the inducer assembly on the collector box and the support bracket to the coupling box.
9. Connect the drain tube from the collector box to the inducer outlet box.
10. Connect the small drain tube from the top of the trap to the fitting on the bottom of the inducer housing.
11. Install the main control box on the blower shelf.
12. Reconnect the edge connector at the main control box on the blower shelf.
13. Reconnect the vent pipe. Check to see that all clamps are tight.
14. Check the condition of the gasket on the cell inlet panel of the burner enclosure. Replace the gasket if necessary (see Figure 2.17).
15. Install the control assembly in the furnace.

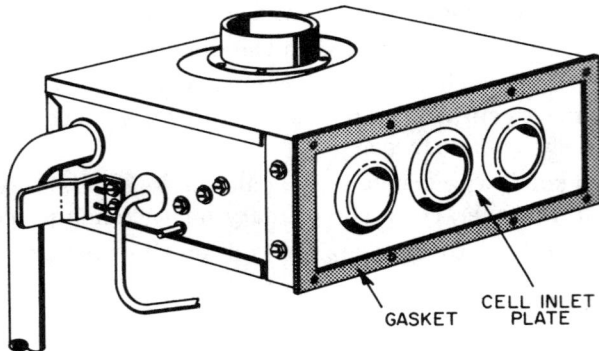

GASKET   CELL INLET
          PLATE

Figure 2.17 Burner enclosure. (Courtesy of Carrier Corporation.)

16. Install the burner enclosure front.
17. Reconnect the pilot leads at the 3-circuit connector.
18. Reconnect the high-voltage lead to the spark generator.
19. Refer to the furnace wiring diagram and connect the wires to the gas valve.
20. Reconnect the pressure tubes to the gas valve and the burner enclosure. Be sure that the tubes are not kinked.
21. Use a second wrench to hold the assembly in place while installing the gas pipe in the gas valve.
22. Reconnect the gas pipe at the ground joint union.
23. Reconnect the combustion-air pipe. Tighten the clamps.
24. Turn on the gas and electrical supplies.
25. Check for gas leaks.

> **Warning:** Never use matches, candles, flame, or other sources of ignition to check for gas leakage. Use a soap-and-water solution, to avoid the possibility of fire, personal injury, or death.

26. Check the furnace operation through two complete operating cycles.
27. Check the pilot tube and gas valve manifold connection for leaks while the furnace is in operation.
28. After the condensate starts to drain, check for condensate leaks.
29. Replace the control and blower access panels.

Pilot

Check the pilot and clean if necessary at the beginning of each heating season. The pilot flame should be high enough for proper impingement on the safety element and to light the main burners. Remove any accumulation of soot and carbon from the safety element. Check the spark electrode gap (see Figure 2.18).

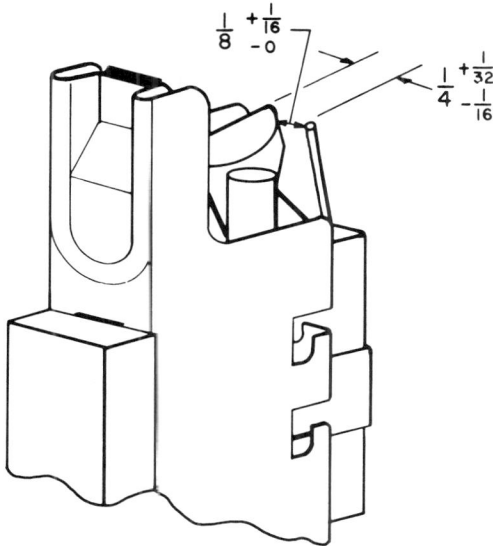

Figure 2.18 Position of electrode to pilot. (Courtesy of Carrier Corporation.)

Electrical Controls and Wiring

Note: The unit may have more than one electrical supply.

With the electrical power disconnected from the unit, check all electrical connections for tightness, and tighten the screws as necessary. If any connections are smoky or burned, disassemble the connection, clean all parts, strip the wire, and reassemble properly and securely. The electrical controls are difficult to check without proper instrumentation; therefore, reconnect the electrical power to the unit and observe the unit through two complete operating cycles.

GAS CONVERSION KIT INSTALLATION INSTRUCTIONS

The following are the instructions for converting the LP (propane) IID (intermittent ignition device) to natural gas IID units:

This instruction covers the installation of the gas conversion kit P/N 308001-708/58 SX-900-021 in a delux condensing gas furnace with LP (propane) controls. The kit may also be used to convert the unit to natural gas

100% shutoff. See Section II for the conversion details. The kit is designed for furnaces with 40 000 through 120 000 Btuh nominal capacity.

Note: Read the entire instruction before starting the installation. There are additional parts shipped in the kit. When the installation is complete, discard the unused parts. The kit contains the following items:

Regulator spring for Model 646 AX-1 Gas Valve   (1)
Pilot orifice   (1)
Main burner orifice   (6)
Pipe plug ($\frac{1}{8}$ in.)   (1)
Installation instructions   (1)

I. *Natural Gas Conversion:* Use the following steps when making this conversion:
   A. Installation of Pilot Orifices:
      1. Turn off the gas and electric supplies to the furnace.
      2. Remove the control access panel.
      3. Remove the burner enclosure front.
      4. Unplug the spark electrode wire from the pilot.
      5. Remove the screws that secure the bottom of the burner enclosure and let it rest on the gas valve.
      6. Using a backup wrench, remove the gas supply tube from the pilot.
      7. Remove and discard the pilot orifice from the gas supply opening of the pilot.
      8. Install the new pilot orifice provided in the kit.
      9. Reinstall the pilot gas supply tube in the pilot. When tightening the pilot gas tube, use a backup wrench.
   B. Installation of Main Burner Orifices:
      1. Remove the screws that secure the burners and remove the burners. For ease of removing the burner, use a 1/4-in. socket and ratchet wrench.
      2. Remove and discard the orifices from the gas manifold.
      3. Install the orifices provided in the kit. Finger tighten the orifices so as not to cross thread them, then tighten them with a wrench. There are enough orifices in each kit for the largest furnace. Discard the extra orifices.
      4. Remove and discard the spoiler screw from each burner (see Figure 2.19).
      5. Reinstall the burners.
      6. Reconnect the spark electrode wire to the pilot.

SPOILER SCREW

BURNER

A85157

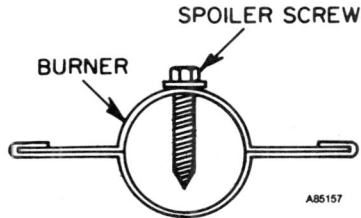

Figure 2.19 Spoiler screw location.
(Courtesy of Carrier Corporation.)

7. Reinstall the burner enclosure bottom.
8. Do not reinstall the burner enclosure front at this time.

C. Conversion of Gas Valve:

1. Remove the regulator seal cap (see Figure 2.20).
2. Remove the adjustment screw and regulator spring.

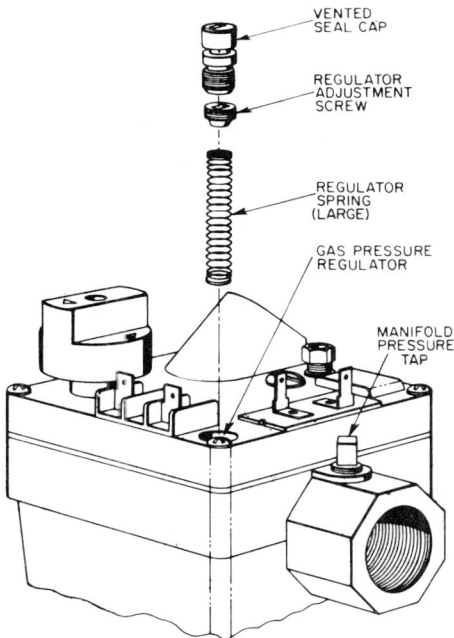

VENTED
SEAL CAP

REGULATOR
ADJUSTMENT
SCREW

REGULATOR
SPRING
(LARGE)

GAS PRESSURE
REGULATOR

MANIFOLD
PRESSURE
TAP

Figure 2.20 Conversion of gas valve.
(Courtesy of Carrier Corporation.)

3. Install the new regulator spring provided in the kit.
4. Replace the regulator adjustment screw.
5. Remove the gas valve conversion plate from the side of the gas valve. No reference to LP (propane) should appear on the furnace.
6. Disconnect the wires from the pressure switch, and remove the pressure switch along with the 1/8-in. pipe nipple from the gas valve.

Note: Use LP (propane) gas-resistant pipe dope. Do not use Teflon tape.

7. Apply the pipe dope sparingly to the 1/8-in. pipe plug provided in the kit and install the plug in the top of the gas valve where the pipe nipple was removed.
8. Remove the furnace notification plate from the left side of the furnace casing alongside of the furnace rating plate (see Figure 2.21).

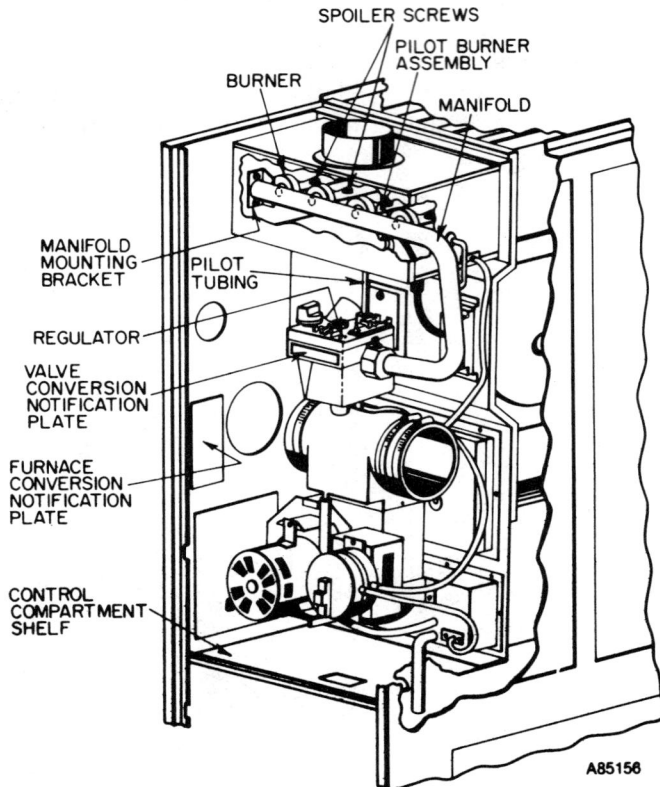

Figure 2.21 Furnace component location. (Courtesy of Carrier Corporation.)

D. Removal of Lockout Timer Module: Use the following steps for the removal of the lockout timer module:
1. Disconnect the long orange wire from terminal 6 on the lockout timer module. Then discard the wire.
2. Disconnect the orange wire at the splice connection located on the right side of the furnace in front of the lockout timer module.

Leave the splice connector on the orange wire from the main control box. Discard the loose orange wire.

3. Disconnect the orange wire from terminal 8 on the lockout timer module, and connect the orange wire from the main control box at the splice connector.

4. Disconnect the blue wire from terminal 7 on the lockout timer module.

5. Disconnect the blue wires from terminal 2 on the gas valve. Remove the blue wire from the piggyback terminal and reconnect it to terminal 2 on the gas valve.

6. Remove and discard the loose blue wire with the piggyback terminal.

7. Disconnect the white wire from terminal 9 on the lockout timer module.

8. Disconnect the white wires from terminal 1 on the gas valve. Remove the white wire from the piggyback terminal and reconnect it to terminal 1 on the gas valve.

9. Remove and discard the loose white wire with the piggyback terminal.

10. Remove the screws securing the lockout timer module to the bracket and remove the module.

11. Remove the screws securing the mounting bracket to the blower shelf and remove the bracket.

E. Check Furnace Operation and Make Necessary Adjustments: Use the following steps to check the furnace operation:

1. Attach a manometer at the pressure tap on the downstream side of the gas valve.

2. Set the room thermostat to "call for heat."

3. Turn on the main gas supply.

Warning: Never use a match or other open flame to check for leaks. Use a soap-and-water solution.

---

4. Turn the manual gas valve to ON and check for gas leaks at the pipe plug in the top of the gas valve.

5. Turn on the electrical supply.

6. When the pilot ignites, check the pilot gas-supply tube connection for leaks. When the main burners ignite, check the manifold orifices for gas leaks.

7. Set the manifold gas pressure at 3.5 in. water column (wc). Clock the input and reset the manifold gas pressure when necessary.

Note: The manifold gas pressure must always be measured with the burner enclosure front removed.

8. Replace the regulator seal cap.
9. Replace the burner enclosure front.

Warning: Be sure that the burner enclosure front is in place after the adjustment has been made.

---

10. Turn the manual gas valve to OFF.
11. Remove the manometer and replace the pressure tap plug.
12. Turn the manual gas valve to ON.
13. With the main burners ignited, check the pressure tap plug for a gas leak.

II. Natural Gas 100% Shutoff Conversion: When natural gas 100% shutoff conversion is desired, follow the conversion procedures in Section I, Parts A–C.

A. Electrical Wiring with Lockout Timer Module in Place:
1. Disconnect the orange wire from terminal 6 of the lockout timer module. Remove and discard the loose orange wire.
2. Disconnect the orange wire at the splice connection located on the right side of the furnace in front of the lockout timer module. Leave the splice connector on the orange wire that was disconnected from the pressure switch. Remove and discard the loose orange wire.
3. Route the orange wire, which is connected at the splice connector, down to the lockout timer module and connect it to terminal 6.

B. Check Lockout Timer Module Operation:
1. Turn off the electrical power supply to the unit.
2. With the furnace off, remove the wire from terminal 5 of the gas valve.
3. Set the room thermostat to "call for heat."
4. Turn on the electrical supply.
5. Let the pilot spark until the lockout module breaks the spark generator circuit (approximately 5 minutes).
6. Replace the wire on terminal 5 of the gas valve.
7. Electrically reset the lockout timer module by setting the room thermostat below room temperature for approximately 30 seconds.
8. Replace the control access panel.
9. Set the room thermostat to the desired temperature.

During normal operation, if the pilot flame is not proven within approximately 5 minutes, the lockout timer module will open, deenergizing the gas valve and stopping the gas flow to the pilot. The lockout timer module will remain open until it is electrically reset.

INSTALLATION INSTRUCTIONS FOR NATURAL-TO-PROPANE GAS CONVERSION KIT (Part Number 58SX900031)

This instruction covers the installation of gas conversion kit 58SX-900–031 in a delux condensing gas furnace with a 646AX–1 gas valve. The kit may also be used for natural gas 100% shutoff. See Section II for conversion details. This kit is designed for use in furnaces with 40 000 through 120 000 Btuh nominal capacity.

> **Note**: Read all of the instructions before starting the installation. There are additional parts shipped in the kit. When the installation is complete, discard the unused parts.

The kit contains the following items:

Regulator spring for Model 646AX-1 gas valve	(1)
Pilot orifice	(1)
Main burner orifice	(6)
Conversion label	(2)
Lockout timer module	(1)
Lockout timer bracket	(1)
Pressure switch	(1)
Pipe nipple	(1)
Wire assemblies	(4)
Screw (spoiler)	(9)
Screw	(2)
Wire tie	(3)
Installation Instructions	(1)

I. LP (Propane) Conversion: Use the following steps when making the conversion:
    A. Installation of Pilot Orifice:
        1. Turn off the gas and electric supplies to the furnace.
        2. Remove the control access door.
        3. Remove the burner enclosure front.
        4. Unplug the spark electrode from the pilot.

    5. Remove the screws that secure the bottom of the burner enclosure and let it rest on the gas valve.

    6. Using a backup wrench, remove the gas supply tube from the pilot.

    7. Remove and discard the pilot orifice from the gas supply opening of the pilot.

    8. Install the new pilot orifice provided in the kit.

    9. Reinstall the pilot gas supply tube on the pilot. When tightening the pilot tube, use a backup wrench.

B. Installation of Main Burner Orifice:

    1. Remove the screws that secure the burners and remove the burners. For ease of removing the burners, use a 1/4-in. socket and ratchet wrench.

    2. Remove and discard the orifices from the gas manifold.

    3. Install the orifices provided in the kit. Finger tighten the orifices so as not to cross thread them, then tighten them with a wrench. There are enough orifices in each kit for the largest furnace. Discard the extra orifices.

    4. Install the spoiler screw in the hole located on the top side of each burner. Be sure to drive the screw in straight (see Figure 2.19).

    5. Reinstall the burners.

    6. Reinstall the burner enclosure bottom.

    7. Reconnect the spark electrode wire to the pilot.

    8. Do not install the burner enclosure front at this time.

C. Conversion of Gas Valve:

    1. Remove the regulator seal cap (see Figure 2.20).

    2. Remove the adjustment screw and regulator spring.

    3. Install the new regulator spring provided in the kit.

    4. Replace the regulator adjustment screw.

    5. Attach the valve conversion plate to the side of the gas valve (see Figure 2.21).

    6. Remove the 1/8-in. pipe plug from the top center of the gas valve.

Note: Use LP (propane) gas-resistant pipe dope. Do not use Teflon tape.

    7. Apply the pipe dope sparingly to both ends of the pipe nipple provided in the kit, and install the nipple in the 1/8-in. tapped opening in the top of the gas valve. Finger tighten.

    8. Install the pressure switch provided in the kit on the pipe nipple. After the switch has been finger-tightened, use pliers on the fitting for the final tightening. When the pressure switch is tight, the terminals should point toward the right side of the furnace (see Figure 2.22).

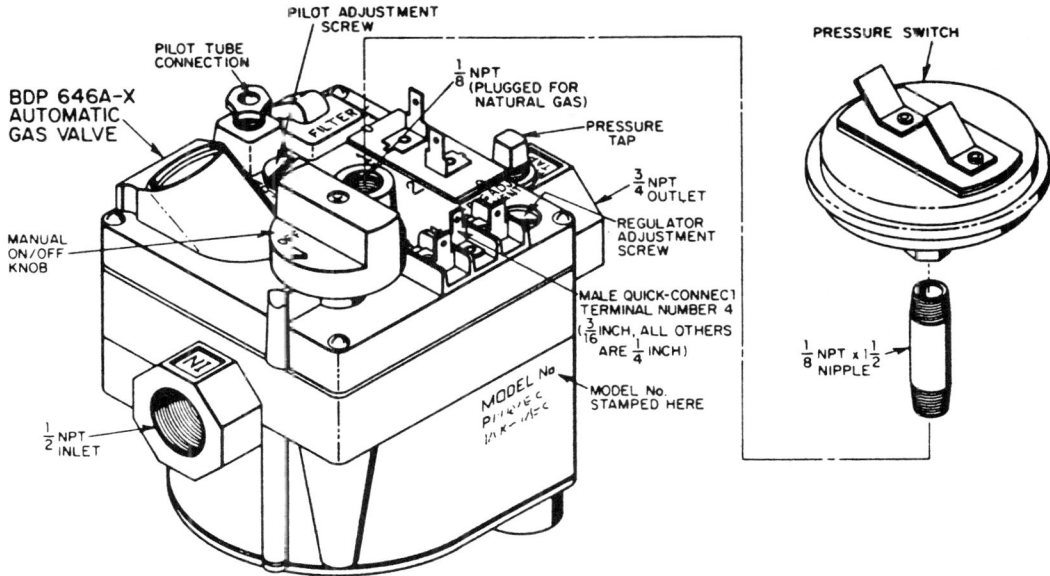

Figure 2.22 Installation of pressure switch. (Courtesy of Carrier Corporation.)

9. Attach the furnace conversion notification plate to the left side of the furnace casing alongside of the furnace rating plate (see Figure 2.21).

D. Installation of Lockout Timer Module:

1. Position the lockout module mounting bracket on the blower shelf, as shown in Figure 2.23.
2. Center punch and drill two 1/8-in. holes in the blower shelf (see Figure 2.23).
3. Install the mounting bracket. (Use the short screws.)
4. Position the lockout module in front of the mounting bracket. Place the lockout module so that the terminals are on the bottom. Do not mount the lockout module to the bracket at this time.
5. Disconnect the orange wire at the splice located on the right side of the furnace in front of the lockout module. Leave the splice connector on the orange wire from the main control box.
6. Connect the short orange wire, provided in the kit, to the splice connector.
7. Route the other end of the short orange wire behind the pressure tube and up the right side of the furnace casing. Connect the orange wire to either terminal on the pressure switch.
8. Connect the long orange wire provided in the kit to the remaining terminal on the pressure switch.

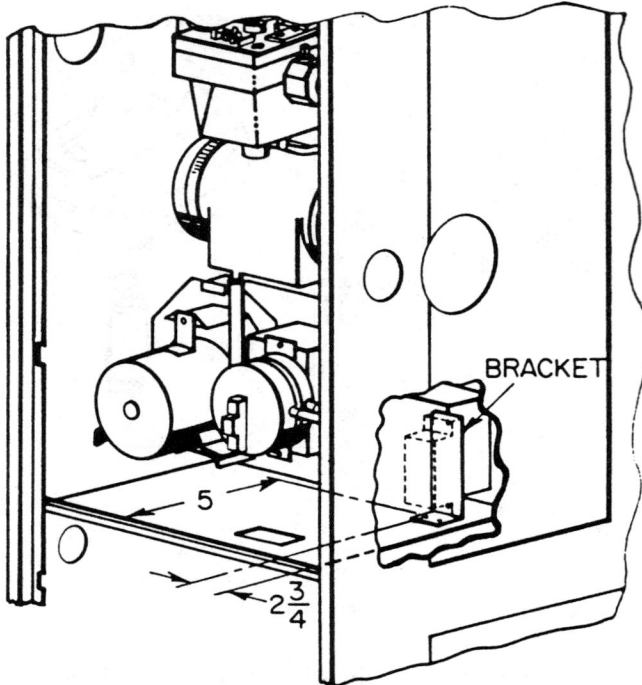

Figure 2.23 Location of timer module (in.). (Courtesy of Carrier Corporation.)

9. Route the other end of the long orange wire down the right side of the furnace casing, behind the pressure tube, and connect it to terminal 6 on the lockout timer module.
10. Carefully cut the wire tie located near the end of the orange wire where the splice connector was removed. Connect the loose orange wire from the inducer control box to terminal 8 on the lockout module.
11. Disconnect the blue wire from terminal 2 on the gas valve.
12. Connect the piggyback terminal of the blue wire provided in the kit to terminal 2 on the gas valve.
13. Reconnect the blue wire to the piggyback on terminal 2 on the gas valve.
14. Route the loose end of the blue wire along the right side of the casing, behind the pressure tube, and connect it to terminal 7 on the lockout module.
15. Disconnect the white wire from terminal 1 on the gas valve.
16. Connect the piggyback terminal of the white wire provided in the kit to terminal 1 on the gas valve.
17. Reconnect the white wire to the piggyback terminal 1 on the gas valve.

18. Route the loose end of the white wire along the right side of the casing, behind the pressure tube, and connect it to terminal 9 on the lockout module.

19. Start one screw in the bottom hole of the mounting bracket. Place the lockout module on the bottom screw.

20. Properly position the lockout module on the bracket and drive in the top screw. Drive the bottom screw the rest of the way in.

21. Using the wire tie provided in the kit, tie the gas valve and pressure switch wires to the gas manifold.

22. Gather the wires together and tie them with the wire tie just above the splice connector.

E. Check Furnace Operation and Make Necessary Adjustments:

1. Attach a manometer at the pressure tap on the downstream side of the gas valve.

2. Set the room thermostat to call for heat.

3. Turn on the main gas valve.

4. Turn the manual gas valve to ON and check the pressure switch connections for gas leaks.

Warning: Never use a match or other open flame to check for gas leaks. Use a soap-and-water solution.

---

5. Turn on the electrical supply.

6. When the pilot ignites, check the pilot gas-supply tube connections for leaks. When the main burners ignite, check the gas manifold orifices for gas leaks.

7. Set the manifold gas pressure to 10.5 in. water column (wc).

Note: The manifold gas pressure must always be measured with the burner enclosure front removed.

8. Replace the regulator seal cap.

9. Replace the burner enclosure front.

Caution: Be sure the burner enclosure front is in place after the adjustment has been made.

---

10. Turn the manual gas valve to OFF.

11. Remove the manometer and replace the pressure tap plug.

12. Turn the manual gas valve to ON.

13. With the main burners ignited, check the pressure tap plug for gas leaks.

F. Check Lockout Timer Module Operation:
1. Turn off the electrical supply.
2. With the furnace off, remove the wire(s) from terminal 5 of the gas valve.
3. Set the room thermostat to call for heat.
4. Turn on the electrical supply.
5. Let the pilot spark until the lockout module breaks the spark generator circuit (approximately 5 minutes).
6. Replace the wire(s) on terminal 5 of the gas valve.
7. Electrically reset the lockout module by setting the room thermostat below room temperature for approximately 30 seconds.
8. Replace the control access panel.
9. Set the room thermostat to the desired temperature.

During normal operation, if the pilot flame is not proven within approximately 5 minutes, the lockout module opens, deenergizing the gas valve, and stopping the gas flow to the pilot. The lockout timer will remain open until it is electrically reset.

G. Check Pressure Switch Operation:
The pressure switch that was installed is a safety device used to guard against possible ignition of unburned gas in the combustion chamber. Gas can accumulate in the combustion chamber immediately after a temporary interruption in the gas supply to the unit; adverse burner operation characteristics can result from low gas-supply pressure.

This normally open switch closes when the gas is supplied to the gas valve under normal operating conditions. The closed switch completes the control circuit through the gas valve.

Should an interruption or reduction in the gas supply occur the gas pressure through the valve drops below the setting of the pressure switch, and the switch will open. Any interruption in the control circuit through the gas valve in which the pressure switch is wired instantly closes the gas valve and stops the gas flow to the burners and to the pilot.

When normal gas pressure to the gas valve is restored and the electric spark ignition system cycles, the unit will return to the normal heating operation.

Before leaving the installation, observe the unit through a few complete cycles of heating operation. During this time, turn off the gas supply to the gas valve just long enough to completely extinguish the burner flame, then instantly restore the full gas supply to the gas valve. To ensure proper pressure switch operation, observe that there is no gas supply to the burners until the electric spark ignition system cycles.

II. Natural Gas 100% Shutoff Conversion:

   Note: Do not use the pressure switch provided in the kit.

   A. Installation of Lockout Module:
      1. Follow steps 1 through 4 of Section I, Part D, in this section.
      2. Remove the orange wire splice connector located on the right hand side of the furnace in front of the lockout module.
      3. Connect the orange wire from the main control box to terminal 6 of the lockout module.
      4. Proceed with steps 10 through 22 of Section I, Part D, to complete the installation of the module.
      5. Check the lockout module operation by following the procedures outlined in Section I, Part F, Steps 2 through 5 of this section.

3

Heil Gas Furnaces

In addition to the information given in Chapter 1, the following information is given for this particular furnace. Even though the information is primarily intended for this manufacturer's furnace some of it may also be used when working on other brands of furnaces.

SAFETY RULES

> **Warning:** Read these rules and all of the instructions given here as well as those given in Chapter 1 of this book carefully. Failure to follow these rules and instructions could cause a malfunction of the furnace, which could result in death, serious bodily injury, and/or property damage.

The information contained here is intended for use by a qualified service technician who is familiar with the safety procedures required and who is equipped with the proper tools and testing instruments.

Installation and repairs made by unqualified persons can result in hazards subjecting that person and others to the risk of injury that can be serious or even fatal.

We will not be responsible for any injury, property, or equipment damage arising from improper installation, service, or repair procedures.

1. Do not install this furnace in a mobile home, trailer, or recreational vehicle.
2. Use only the type of gas approved for this furnace (see the rating plate). Overfiring will result in failure of the heat exchanger and cause dangerous operation.
3. This furnace must be connected only to an approved vent system to carry combustion products outdoors as described in Chapter 1 of this book.
4. Never test for gas leaks with an open flame. Use soap suds to check all connections. This will help to avoid any possibility of fire or explosion.
5. Provide adequate combustion and ventilation air to the furnace area. Refer to Chapter 1 of this book.
6. Make sure the supply and return ducts are sealed to the furnace casing and entirely separate from the area supplying the combustion and ventilation air.

> **Note**: It is the responsibility and obligation of the customer to contact a qualified installer to assure that the installation is adequate and is in conformance with the governing codes and ordinances.

INSTALLATION REQUIREMENTS

The installation of this furnace must conform with local building codes or in the absence of local codes, with the American National Standards, Z223.1, 1984, National Fuel Gas Code and the National Electrical Code, ANSI/FPA 70, 1984, or current editions.

A typical installation is shown in Figure 3.1. It shows the basic connecting parts needed to install the furnace. In addition to these parts, supply and return plenums and ducts are needed.

BTU INPUT RATING

The Btuh input rating of these furnaces can be changed from the standard rating to the alternate input rating as listed on the rating plate. To change the input rating the main burner orifices must be changed. Refer to the Tables 3.3 and 3.4 for the proper size orifices. Changing the orifices must be done by a qualified service technician, preferably before connecting the gas supply. See Changing Main Burner Orifices in the Furnace Maintenance Section of this chapter.

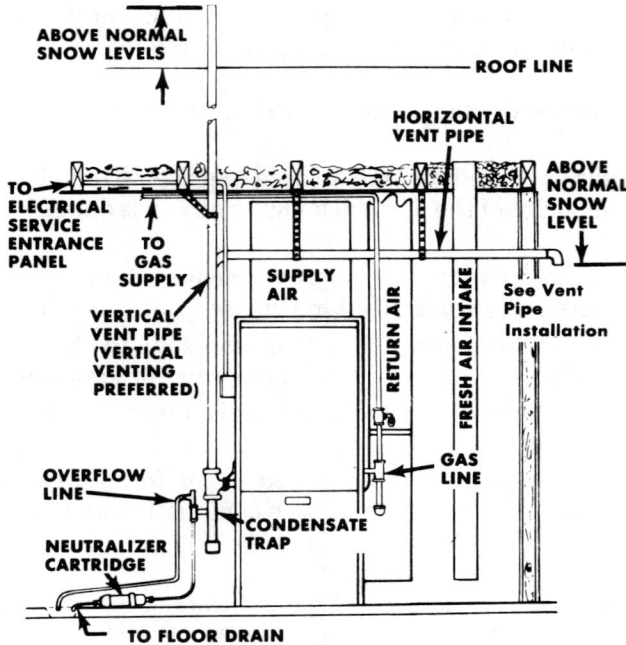

Figure 3.1 Typical installation. (Courtesy of Heil-Quaker Corporation.)

START-UP PROCEDURE

On a new installation or if a major part such as the gas valve, pressure switch, or fan/limit control, has been replaced the operation of the furnace must be checked.

Check the furnace operation as outlined in the following instructions. If any sparking, odors, or unusual noises are encountered, shut off the electric power immediately. Recheck for wiring errors, or obstructions in or near the blower motors.

> **Warning**: Danger of explosion or fire. Liquified petroleum (LP) gas is heavier than air and it will settle in any low area, including open depressions and it will remain there unless the area is ventilated.

Never attempt the start-up of a unit before thoroughly ventilating the area.

Start the Furnace

Start the furnace according to the instructions given with the furnace and perform the following checks and adjustments.

Check the gas input and pressures.   For furnaces located at altitudes between sea level and 2000 ft, the measured input must not be greater than the input shown on the rating plate of the furnace. For elevations above 2000 ft, the measured input must not exceed the input on the rating plate reduced by 4% for each 1000 ft that the furnace is above sea level.

The gas supply pressure and manifold pressure with the burners operating must also be as specified on the rating plate (see Table 3.1). The rated input will be obtained when 2500 Btu propane at 10 in. water column (wc) manifold pressure with the factory sized orifices. If LP gas with a different Btu rating is used, the orifices must be changed by a licensed petroleum gas installer before the furnace is operated.

TABLE 3.1 FURNACE MANIFOLD GAS PRESSURES
(COURTESY OF HEIL-QUAKER CORPORATION)

Type of gas	Manifold pressure, in. W.C.
Natural	3.5
L.P.	10.0

Check the manifold gas pressure.   A tapped opening is provided in the gas valve to aid in measuring the manifold gas pressure. A U-tube manometer having a scale range from 0 to 12 in. of water column should be used for making this measurement (see Figure 3.2). The manifold pressure must be measured with the main burner and the pilot burner operating.

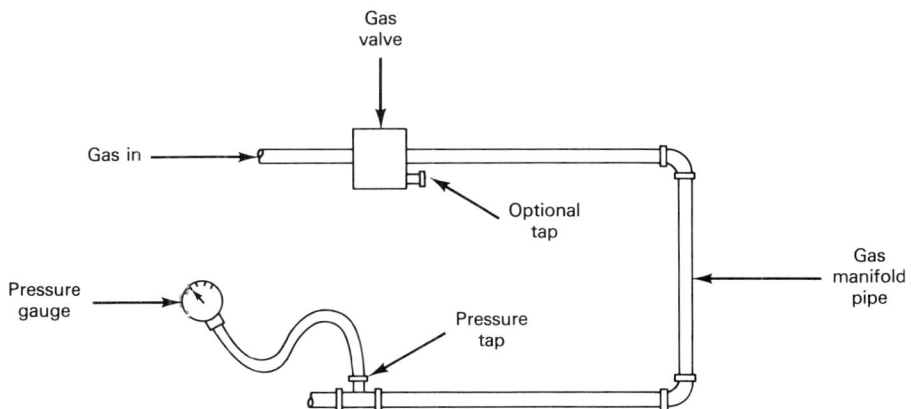

Figure 3.2 Measuring manifold gas pressure.

To adjust the gas pressure regulator, remove the adjustment screw or cover on top of the gas valve (labeled HI on some valves) and turn the screw out (counterclockwise) to decrease the gas pressure or turn the screw in (clockwise) to increase the gas pressure. Only small variations in the gas flow should be made by means of the gas pressure regulator adjustment. In no case should the final manifold gas pressure vary more than plus or minus 0.3 in. water column from the above specified pressures. Any major changes in the gas flow should be made by changing the size of the burner orifice.

Check the gas input (natural gas only).   To measure the input using the gas meter proceed as follows:

1. Turn off the gas supply to all appliances except the furnace.
2. With the furnace operating, time the smallest dial on the meter for one complete revolution. If this is a 2-ft³ dial, divide the seconds by 2; if it is a 1-ft³ dial, use the time in seconds per cubic foot of gas being delivered to the furnace.
3. Example: Natural gas with a heating value of 1000 Btu per cubic foot and 34 seconds per cubic foot as determined by step 2; then

$$\text{Input} = 1000 \times 3600 \div 34$$
$$= 106\,000 \text{ Btu per hour}$$

Note: The Btu content of the gas should be obtained from the supplier. This measured input must not be greater than the input indicated on the rating plate of the furnace.

4. Relight all of the other appliances that were turned off in step 1 above. Be sure that all pilot burners are operating.

PRIMARY AIR ADJUSTMENT

If the burners are not equipped with air shutters, no adjustment is necessary. Adjustment of the air shutter may be necessary to obtain the correct flame characteristics and/or to minimize resonance heat exchanger noise that is generated by the main burner flame.

a. Check the air shutter position; it should be full open.
b. Start the furnace; see the lighting instructions on the furnace or in the owners information manual.
c. Allow the furnace to run for 10 minutes, then check the flame characteristics (see Check Main Burner Flame later in this chapter).
d. To adjust, loosen the shutter locking screw(s) and close the shutter until

the flame has a yellow tip and then open just enough to eliminate the yellow tip, tighten the locking screw(s).

e. If a resonance noise occurs, close the air shutters just enough to permit the slightest amount possible of yellow tip in the flame; tighten the locking screws.

> Note: A burner spoiler screw may be added for extreme cases of resonance noise that cannot be held to an acceptable level by air shutter adjustment.

f. Install a number 8 × 1-in.-long sheet metal screw in the side of the burner. A pilot hole for the screw is already in the burner. Install the screw only far enough into the burner to be effective in conjunction with the air shutter to obtain acceptable flame characteristics and noise level. Repeat the adjustment as necessary using a combination of spoiler screw and air shutter to obtain the proper flame characteristics and an acceptable noise level. Check to ensure that the adjustment has not caused any sooting of the heat exchanger. Tighten the air shutter locking screws.

g. Apply high-temperature silicone sealant (500 °F) to the spoiler screw/ burner to retain the adjusted position.

ADJUST PILOT BURNER

The furnace has an intermittent pilot. The flames should encase three-eighths to one-half in. of the ignitor sensor tip (see Figure 3.3). To adjust, remove the cap or screw from the pilot adjusting screw on the gas valve (see Figure 3.11). Turn the screw counterclockwise to increase or clockwise to decrease the flame as required. Replace the cap on the adjusting screw. The pilot flame should be inspected monthly.

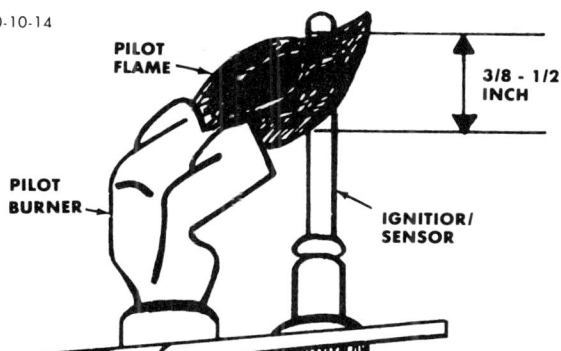

Figure 3.3 Pilot flame adjustment. (Courtesy of Heil-Quaker Corporation.)

CHECK LIMIT AND FAN CONTROL

Check the limit control function after 15 minutes of furnace operation by blocking the return air grill(s).

1. After several minutes the main burners must go off. The blower will continue to run.
2. Remove the air restrictions and the main burner will relight after a cool down period of a few minutes.

Adjust the room thermostat setting below the room temperature.

1. The main burners must go off.
2. The circulating air blower should continue to run briefly until the supply air temperature drops to approximately 100 to 90 °F.

The fan and limit controls are preset at the factory. The control is set for the fan to go off at 100 to 90 °F (see Figure 3.4).

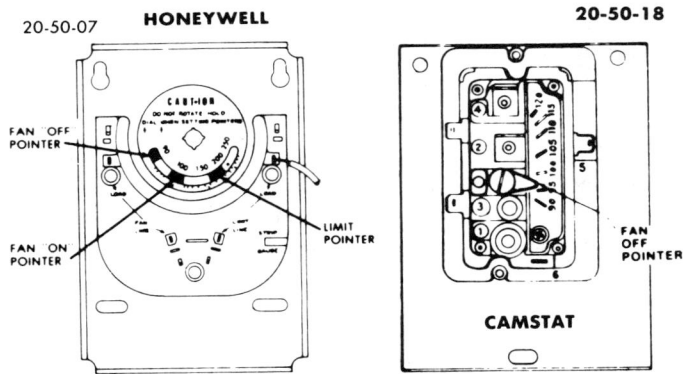

Figure 3.4 Fan and limit control. (Courtesy of Heil-Quaker Corporation.)

Note: If necessary, adjust the fan off setting to obtain a satisfactory comfort level.

Warning: *Danger of fire: The limit control is factory preset and must not be adjusted.*

CHECK TEMPERATURE RISE

Check the temperature rise through the furnace by placing thermometers in the supply and return air registers as close to the furnace as possible.

1. All registers and duct dampers must be open and the unit should have operated for 15 minutes before taking these readings.
2. It must be within the range specified on the rating plate of the furnace.

> Note: The air temperature rise is the difference between the supply air and the return air temperatures.

With a properly designed system, the proper amount of temperature rise will normally be obtained when the unit is operating at the rated input with the recommended blower speed.

If the correct amount of temperature rise is not obtained, it may be necessary to change the blower speed. A higher blower speed will lower the temperature rise. A slower blower speed will increase the temperature rise.

> Note: The blower speed must be set to give the correct air temperature rise through the furnace as marked on the furnace rating plate.

CHANGING BLOWER SPEEDS

See Table 3.2 for the proper blower performance chart for the furnace in question.

> Warning: Danger of electrical shock, bodily injury, or death if electric power is not turned off before changing speed taps.

If it is necessary to change the circulating air blower speed, the terminal block in the furnace junction box makes this a simple operation (see Figure 3.5). The yellow wire is connected to the speed tap for heating. The violet wire is connected to the HI speed tap for cooling. Unplug and connect them to the desired speed tap.

If you must use the same speed tap for both heating and cooling, install a duplex spade terminal adaptor on the motor speed tap to connect both the yellow and violet wires or strip the yellow lead to expose a bare wire about one and one-half in. back from the terminal. Remove the terminal from the violet lead and strip three-quarters in. from the end. Twist this wire around the ex-

TABLE 3.2 BLOWER PERFORMANCE DATA, ALL MODELS (COURTESY OF HEIL-QUAKER CORPORATION)

			NUGK040CF NUGK050LF NULK050LF	NUGK075AG NUGK075CG NULK075AG	NUGK100AH NUGK100CH NULK100AH	NUGK125AK NUGK125CK NULK125AK
Model number						
Blower data type & size			DD10-8A-T	DD10-9A-T	DD10-9A-T	DD12-11A-T
Motor data	Amps @ RPM		8.0 @ 1050	10.8 @ 1050	10.8 @ 1050	11.1 @ 1050
	Type/hp		PSC-1/2 HP	PSC-3/4 HP	PSC-3/4 HP	PSC-3/4 HP
Capacitor MFD/volts			7.5 MFD/370	10 MFD/370	10 MFD/370	15 MFD/370
Filter data	# Req'd/size		1-16 × 25 × 1	1-16 × 25 × 1	2-16 × 25 × 1	1-24 × 25 × 1
	Type		Permanent	Permanent	Permanent	Permanent
Air delivery in CFM varying ext. static press. (in. WC.)	ESP	Speed				
	0.10	LO	800	1035	1050	1460
		MED. LO	1080	1305	1355	1620
		MED. HI	1350	1545	1660	1950
		HI	1570	1720	1885	2320
	0.20	LO	850	1030	1040	1455
		MED. LO	1075	1270	1330	1615
		MED. HI	1320	1490	1600	1910
		HI	1510	1650	1820	2255
	0.30	LO	855	1020	1035	1460
		MED. LO	1060	1235	1305	1610
		MED. HI	1280	1430	1545	1885
		HI	1445	1580	1750	2205
	0.40	LO	855	995	1025	1445
		MED. LO	1040	1195	1280	1590
		MED. HI	1230	1375	1490	1855
		HI	1375	1510	1675	2145
	0.50	LO	835	960	1010	1430
		MED. LO	1010	1145	1230	1570
		MED. HI	1180	1300	1425	1825
		HI	1300	1435	1585	2105

posed yellow wire several times. Solder and tape. Connect this connection to the speed tap desired.

CHECK MAIN BURNER FLAME

The flames should be soft, stable, and blue (dust may cause orange tips or they may have wisps of yellow, but they must not have solid yellow tips). They

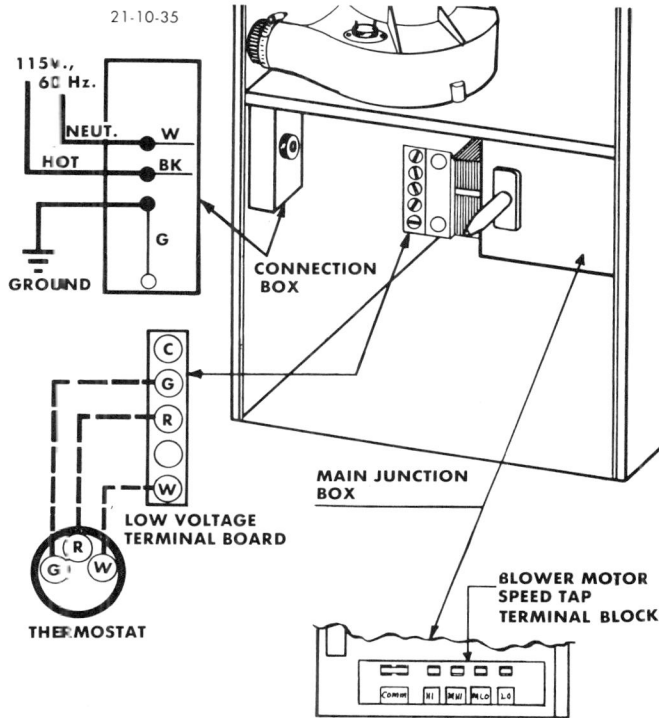

Figure 3.5 Electrical connection. (Courtesy of Heil-Quaker Corporation.)

should extend directly upward from the burner without curling, floating, or lifting off the burner. They must not touch the sides of the heat exchanger (see Figure 3.6). The main burner flame should be inspected monthly.

Main Burner Flame Adjustment

The following procedures are to be used for adjusting main burners. The first one is for natural gas and the second one is for LP gas burners.

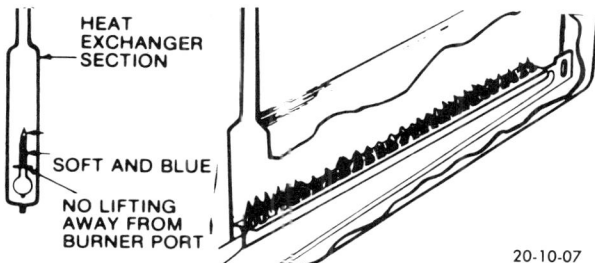

Figure 3.6 Main burner flame. (Courtesy of Heil-Quaker Corporation.)

Natural gas burners only (normal combustion).   Turn the furnace on at the wall thermostat. Wait a few minutes, since any dislodged dust will alter the normal flame appearance. The flames should be stable, quiet, soft, and blue with slightly orange tips. They should not be yellow and they should extend directly upward from the burner ports without curling downward, floating, or lifting off the ports. They should not touch the sides of the heat exchanger.

The causes for abnormal combustion are:

1. *Gas pressure*: Adjust it to 3.5 in. water column.
2. *Gas flow.* Clock the gas meter. The gas flow must be within 5% of the Btu rating of the furnace being tested.
3. *Combustion air.* Check for sufficient combustion air, especially if the furnace is installed in a confined area, such as a closet.
4. *Venting.* Check for blockage or obstructions in the vent system.
5. *Burners.* Check the burners for dirt, soot, rust, and other obstructions.
6. *Flue baffle.* Check the flue baffle for blockage or distortion.
7. *Heat exchanger.* Check the heat exchanger for a burn-out or cracks.

LP gas only (primary air adjustment).   If the burners on the furnace are not equipped with air shutters, no adjustment is necessary. LP furnaces have air adjustments on their burners. Use the following procedures for making the adjustment:

1. Start the main burners as outlined above.
2. Observe the flame characteristics after 10 minutes operation.
3. Check the gas pressure. Adjust the gas pressure to 10 in. water column.
4. If adjustment is necessary, loosen the primary air shutter locking screw. Adjust the air shutter opening to a position that gives a slightly yellow tip in the flame. Open the air shutter slightly to just eliminate the yellow tip in the flame. Tighten the shutter screw (see Figure 3.7).
5. Orifices. Check the orifices for the correct size for an LP gas application.

Spud Orifices.   Use the following procedure to check the spud orifice size:

1. Check the orifice size to make certain that they are sized for the proper application, either for natural gas or for LP gas (see Tables 3.3 and 3.4).
2. Before changing orifices except for obvious damage use the following procedure:
   A. Check and adjust the manifold gas pressure.
   B. Check the gas input to the furnace.

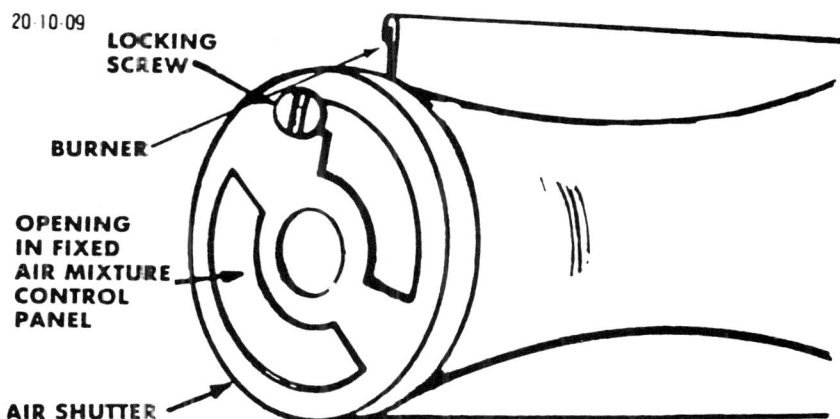

Figure 3.7 Shutter adjustment. (Courtesy of Heil-Quaker Corporation.)

TABLE 3.3 NATURAL GAS ORIFICE GUIDE (COURTESY OF HEIL-QUAKER CORPORATION)

	Nat. gas: 1000 BTU/cu. ft.		1000 BTU/cu. ft.		1000 BTU/cu. ft.	
Drill	3″		3 1/2″		4″	
No.	Cu. ft./hr	BTU./hr	Cu. ft./hr	BTU/hr.	Cu. ft./hr	BTU/hr
50	12	12,000	13	13,000	14	14,000
49	14	14,000	15	15,000	16	16,000
48	15	15,000	16	16,000	17	17,000
47	16	16,000	17	17,000	18	18,000
46	17	17,000	18	18,000	19	19,000
45	18	18,000	19	19,000	20	20,000
44	19	19,000	21	21,000	22	22,000
43	20	20,000	22	22,000	23	23,000
42	22	22,000	24	24,000	26	26,000
41	24	24,000	25	25,000	27	27,000
40	25	25,000	26	26,000	28	28,000
39	26	26,000	28	28,000	30	30,000
38	27	27,000	29	29,000	31	31,000
37	28	28,000	30	30,000	32	32,000
36	30	30,000	32	32,000	34	34,000
35	31	31,000	33	33,000	36	36,000
34	32	32,000	34	34,000	37	37,000

TABLE 3.4 LP GAS ORIFICE GUIDE
(PROPANE) (COURTESY OF HEIL-QUAKER
CORPORATION)

10″ pressure - inches water column	
Propane - 2500 BTU/cu. ft.	
Drill size	BTU/hr.@ 10
41	66767
42	63355
43	57361
44	53580
45	48692
46	47493
47	44634
48	41868
49	38640
50	35505
51	32553
52	29234
53	25637
54	21948
55	19550
56	15659
57	13390
58	12772
59	12173
60	11592

C. Check the primary air shutter adjustment (LP gas only) or other causes listed in the main burner flame adjustment section.

SETTING THE THERMOSTAT HEAT ANTICIPATOR

Visually check the heat anticipator for damage. Check the continuity of the heat anticipator. If there is no continuity the anticipator circuit is open and the thermostat must be replaced.

Amping Out the Anticipator Setting

Turn off the electrical power to the system and remove the thermostat from the subbase or wall plate.

Wrap exactly 10 turns of wire around the jaw of a split-jaw induction-type current meter (see Figure 3.8).

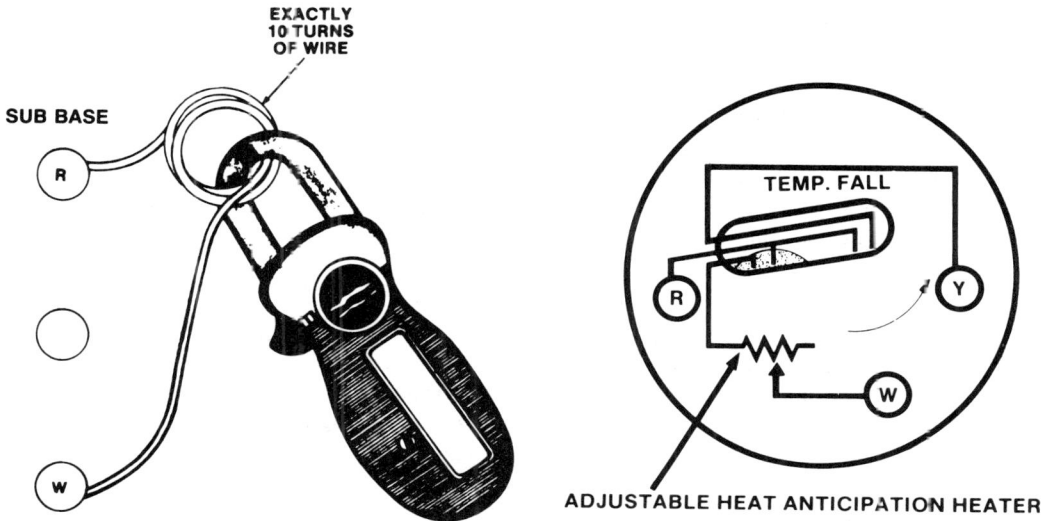

Figure 3.8 Setting thermostat anticipator. (Courtesy of Heil-Quaker Corporation.)

Connect one end of the wire to the R terminal and the other end to the W terminal on the thermostat.

Turn on the electrical power to the system; the furnace should fire. Wait approximately one minute and read the amp scale on the meter. Divide the amperage reading by 10 to obtain the current draw.

Example

The circuit is drawing 2 A. Divide the 2 A indicated on the meter by 10 to obtain the actual amperage draw of the circuit: $2 \div 10 = 0.2$ A.

Turn off the electrical power to the system and remove the amp-meter and wire coil. Set the heat anticipator to match the amperage draw. Mount the thermostat.

The furnace should be inspected and serviced on an annual basis (before the heating season) by a qualified service technician.

It should be remembered that certain mechanical and electrical skills and tools are required to properly perform maintenance on the furnace. Personal injury or death may result if you are not properly trained.

Warning: *Turn off the electrical power to the furnace before performing any maintenance or removing panels, because of the danger of electrical shock.*

Air Filters—Check Monthly

The air filter(s) should be inspected at least monthly and cleaned or replaced as required. There are two types of filters that are most commonly used. The most widely used is the fiberglass disposable type, which should be replaced before it becomes clogged. The other type commonly used is the washable type constructed of aluminum mesh, foam, or reinforced fibers. Washable-type filters may be cleaned by soaking in a mild detergent and rinsing with clear water.

> **Note:** Some filters are marked with an arrow to indicate the proper direction of air flow through the filter. When installing, the arrow must point in the direction of the air flow. Remember that dirty filters are the most common cause of inadequate heating or cooling performance.

Table 3.5 lists the recommended sizes and types of filters that may be used with different furnaces based on air flow through the furnace.

TABLE 3.5 RECOMMENDED FILTER SIZES MINIMUM SQUARE INCHES/NOMINAL SIZE FILTER (COURTESY OF HEIL-QUAKER CORPORATION)

CFM airflow	Disposable type filter low velocity/300 FPM		Cleanable type filter high velocity/500 FPM	
	Minimum surface area (sq. in.)	Recommended nominal size	Minimum surface area (sq. in.)	Recommended nominal size
800	384	20 × 25	231	14 × 20
900	432	20 × 25	260	15 × 20
1000	480	20 × 30	288	14 × 25
1100	528	20 × 30	317	15 × 25
1200	576	14 × 25 (2)	346	16 × 25
1300	624	14 × 25 (2)	375	20 × 25
1400	672	16 × 25 (2)	404	20 × 25
1500	720	16 × 25 (2)	432	20 × 25
1600	768	20 × 25 (2)	461	20 × 25
1700	816	20 × 25 (2)	490	20 × 30
1800	864	20 × 25 (2)	519	20 × 30
1900	912	20 × 30 (2)	548	24 × 25
2000	960	20 × 30 (2)	576	24 × 25

(2) Two Required

In some installations, a larger filter may have been used for additional air volume, or if the furnace was installed for heating only with a remote filter cabinet or central return a smaller filter may have been used. If air conditioning has been added since the furnace was installed, make sure that the filter size is adequate.

Replacement filters should be the same size and type to ensure adequate air flow and filtering, unless a disposable low-velocity filter is replaced with a washable high-velocity type.

Filter replacement.   The filter will normally be found inside the furnace blower compartment, but some alternative locations may be a remote filter rack attached to the outside of the furnace, a separate return air cabinet attached to the furnace, or a remote filter grill.

Remote filter grills and return air cabinets will usually have a hinged door or removable panel that permits removing the filter. Filter racks attached to the furnace will usually be made so that the filter simply slides out one side for removal. Use only the same size filter. The type must be the same unless replacing a disposable low-velocity type with a washable high-velocity type.

> **Warning:** *Never operate a furnace without a filter installed because dust and lint will build up on internal parts, resulting in a loss of efficiency, equipment damage, and possible fire.*

Filter Replacement.   The following steps should be used when replacing a filter:

1. Turn off the electric power to the furnace with a circuit breaker or a disconnect switch.
2. Remove the blower compartment door.
3. Lift the filter strap from the hook on the side or slide the strap to the side from underneath the flange on the front, as shown in Figure 3.9. Remove the filter, being careful not to dislodge dirt and debris.
4. Inspect the filter and replace or clean the washable type. If the filter is aluminum mesh it should be recoated with a filter coating spray.
5. Reinstall the filter under the strap. If the filter is marked for air flow direction, make sure it is installed correctly.
6. Replace the blower compartment door making sure that it is tightly closed.
7. Turn on the electrical power to the furnace.

Blower Motor

The blower motor will require lubrication every five (5) years of normal operation. Add one-half teaspoon (2 ml) of SAE No. 10 W 30 motor oil to each motor bearing through the oil tubes or by removing the cap plugs in the motor end bells.

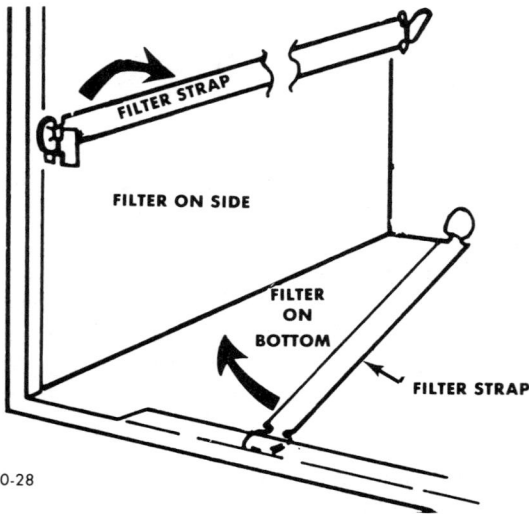

21-10-28

Figure 3.9 Filter replacement. (Courtesy of Heil-Quaker Corporation.)

> **Caution:** *Do not* over oil or use 3 in 1 oil, penetrating oil, WD 40, or similar oils to oil the motor bearings. Use of these oils may damage these motors.

Condensate Disposal (Monthly Disposal)

Condensing gas furnaces have a condensate trap as a part of the vent system. The moisture in the flue gases will condense and collect in the trap, then go to an inside drain or be pumped to a sewer line using a condensate pump.

The condensate trap and condensate neutralizer cartridge (if used) in the drain line leading from the trap will require some maintenance. Disassemble and clean the trap and cartridge prior to each heating season or if the drain line becomes plugged.

Inspect the drain line and the overflow line at least monthly. If the condensate neutralizer cartridge becomes plugged the condensate will flow through the overflow line. If this happens, clean both the cartridge and the trap.

TO CLEAN

Disconnect the drain line cartridge and unscrew the end cap from the cartridge. Pour the neutralizer out and thoroughly flush the neutralizer and the inside of the cartridge with clear water. Pour the neutralizer back into the cartridge if the cartridge is less than three-quarters full. Unscrew the top from the vent connecting tee and flush it thoroughly with water. Use soap if necessary to clean. *Do not use* any kind of solvents. Make sure that the float is reinstalled in the trap (see Figure 3.10).

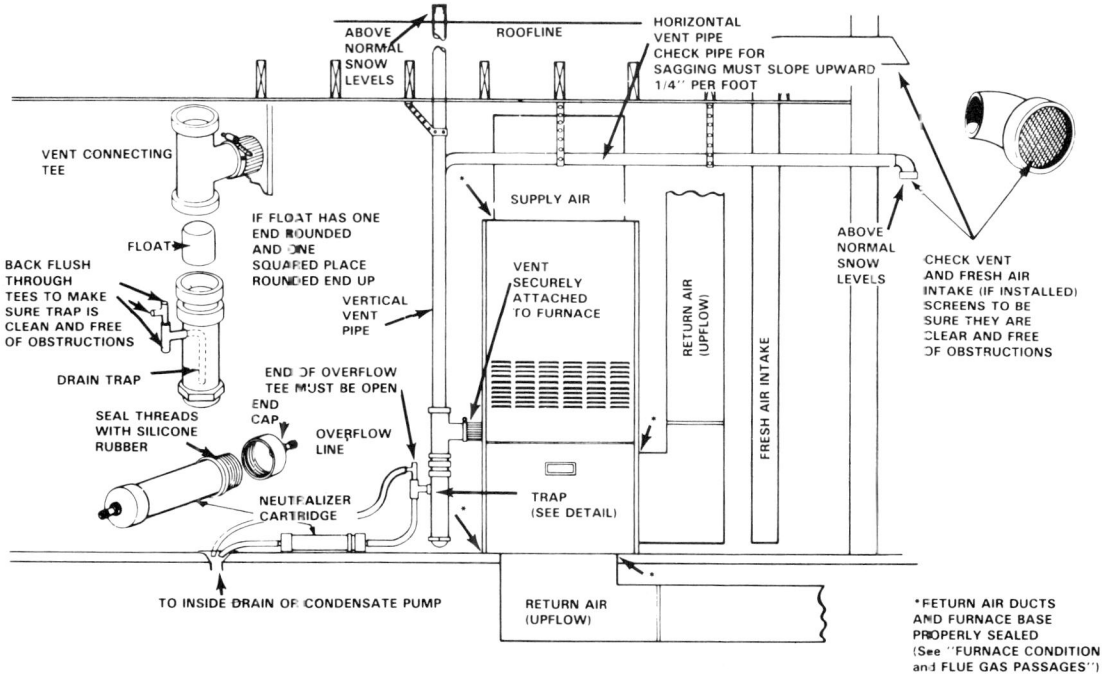

Figure 3.10 Typical furnace installation. (Courtesy of Heil-Quaker Corporation.)

Reassemble and seal the threaded connections with silicone rubber (bathtub calk) or pipe dope approved for plastic pipe.

Do not use the condensate for any reason, because it is acidic.

Furnace Condition and Flue Gas Passages/ Monthly

A properly adjusted gas furnace should not require cleaning at frequent intervals, but it should be inspected regularly to ensure safe and efficient operation. A brief monthly inspection is recommended that does not require disassembly. In addition the furnace should be inspected and cleaned annually (only if required) by a qualified technician.

During the monthly inspection check the vent pipe and fresh air intake (if installed) to be sure they are clear and free of obstructions. Check the vent pipe for evidence of condensate leakage, tight joints, secure attachment to the furnace, and sagging pipe.

Horizontal sections of the vent pipe must slope upward 1/4 in. per foot except the section between the furnace and the drain trap when the trap is not mounted directly on the furnace. That section (maximum of 4 ft long) should slope downward one-fourth in. per foot (maximum drop 1 in.) to trap.

Check the return air ducts to make sure that they are sealed to the furnace casing and that it is in good physical condition. It must terminate outside the space containing the furnace with no holes or inlets into the furnace space.

The floor or furnace base must be in good physical condition. For an upflow furnace with a bottom return, the floor or base area around the furnace must form a seal (no sagging, cracks, defects, etc.) to prevent air from being pulled in from the furnace area, or any defect area must be sealed between the floor or base and furnace.

Remove the front panel and use a flashlight to inspect the visible part of the heat exchanger, burners, and spark ignitor. Check for loose soot and give particular attention to obvious deterioration from corrosion or other sources. Check for any signs of condensate leakage inside the furnace cabinet.

If soot or deterioration is found or if there is evidence of condensate leakage inside the furnace: *Do not operate the furnace until the cause has been found and corrected and the deterioration has been corrected.*

Main Burner and Pilot Flames/Monthly

Allow the furnace to run approximately 10 minutes, then inspect the main burner flames and the pilot flame.

Main burner flames.   The main burner flames should be stable, soft, and blue (dust may cause orange tips or they may have wisps of yellow, but there must not be solid yellow tips). They should extend directly upward from the burner without curling, floating, or lifting off the burner. They must not touch the sides of the heat exchanger.

Pilot flame.   The pilot flame should surround three-eighths to one-half in. of the ignitor/sensor tip (see Figure 3.4).

Changing Main Burner Orifices

This operation is to be performed by a qualified technician only. The main burner orifices can be changed to operate the furnace at an alternative input rating or if required for high altitudes.

> **Warning:** *Danger of property damage, bodily injury, or death if electric power (at disconnect) and gas supply (at manual shutoff valve in gas line) are not shut off.*

Use the following steps when changing the gas orifices in a gas furnace.

1. Disconnect the gas line from the gas valve. The wiring to the gas valve can be disconnected if desired.
2. Disconnect the pilot tube at the gas valve (see Figure 3.11).

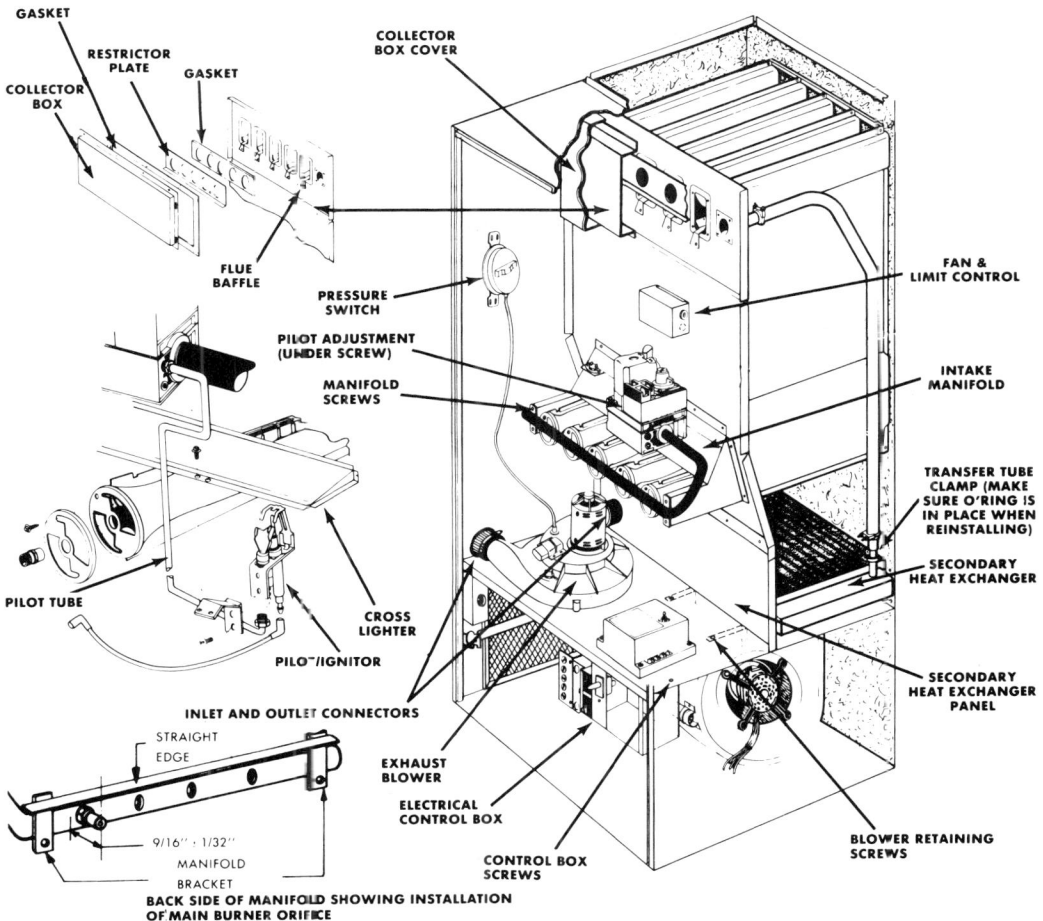

Figure 3.11 Furnace component location. (Courtesy of Heil-Quaker Corporation.)

3. Remove the four screws holding the manifold to the manifold brackets and remove.
4. Remove the orifices from the manifold and replace them with the properly sized orifices.
   a. Tighten the orifices so there is nine-sixteenths in. from the face of the orifice to the face of the manifold brackets. To check, place a straight edge across the manifold brackets and measure to the face of the orifice, as shown in Figure 3.11. Make sure that the orifice goes in straight so that it forms a right angle (90°) to the manifold brackets.
5. Reinstall the manifold and the other parts. Use caution when installing the pilot line to avoid cross-threading or stripping of the threads. Make sure that the burners do not bind on the new orifices.

Cleaning Heat Exchangers (Qualified Service Technician Only)

Furnaces that are properly installed and maintained will normally not require cleaning of the heat exchangers.

If the filters are inadequate or not maintained it may be necessary to clean the exterior surface of the secondary heat exchanger to obtain a proper air flow. The primary heat exchanger can be cleaned without removing or cleaning the secondary heat exchanger.

The only time it should be necessary to disassemble and clean the interior of both primary and secondary heat exchangers would be in the case of a sooting condition caused by abnormal combustion.

> **Warning:** *Danger of property damage, bodily injury, or death if electric power (at the disconnect) and the gas supply (at the manual shutoff valve in the gas line) are not shut off.*

Secondary Heat Exchanger (Exterior Cleaning)

Use the following steps during this procedure:

1. Remove the two retaining screws holding the control box from the top of the blower deck (see Figure 3.11).
2. Lift the control box up and outward from the furnace. The control box will have enough slack in the wiring to allow the box to be held out of the way to remove the blower assembly.
3. Remove the two retaining screws holding the blower in position in the side rails (see Figure 3.11).

4. The blower can now be removed by pulling the assembly from the furnace. Support the blower next to the furnace to avoid having to disconnect the wiring.

5. Using a stiff bristle brush and a vacuum cleaner, clean the dirt and lint buildup from the bottom side of the secondary heat exchanger. The brush strokes must be with the fin surface to avoid damage to the fins. Use a fin comb to straighten the fins (see Figure 3.12).

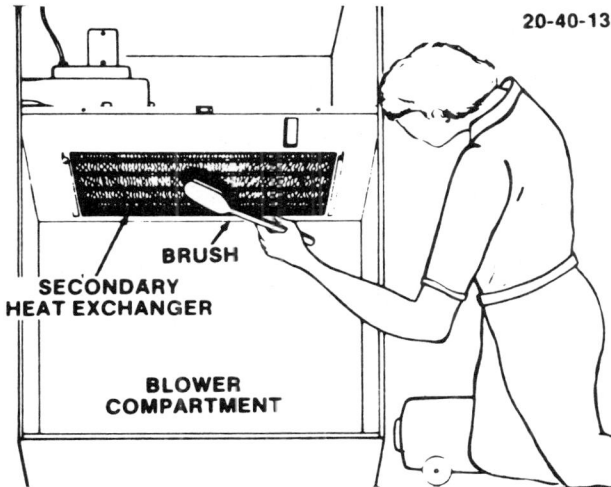

20-40-13

BRUSH
SECONDARY
HEAT EXCHANGER

BLOWER
COMPARTMENT

Figure 3.12 Cleaning the secondary heat exchanger. (Courtesy of Heil-Quaker Corporation.)

6. Inspect and clean the blower wheel using a brush and a vacuum. Be careful not to dislodge the balance weights (clips) that may be on the blower wheel.

Caution: An unbalanced blower wheel may cause undesirable noise, vibration, or blower damage.

Primary Heat Exchanger Disassembly

The following parts and assemblies must be removed before the heat exchanger can be cleaned (refer to Figure 3.11).

1. Remove the screws from the collector box cover, and the collector box and restrictor plate. Handle the collector box gasket with care to avoid damage.

2. Remove the screws that hold the flue baffles in position and carefully pull the baffles out.

3. Disconnect the gas supply line at the union and at the gas valve if necessary for removal from the furnace.

4. Disconnect the electrical lead at the gas valve, ignitor/sensor and the ground wire from the pilot assembly.

5. Disconnect the pilot gas tube from the gas valve.

6. Remove the retaining screws for the manifold and remove the gas valve/manifold assembly.

7. Pull the burners and crosslighter from the heat exchanger.

8. Remove the crosslighter from the burners by sliding the crosslighter toward the orifice end of the burner.

Cleaning.   Use the following steps in the cleaning process:

1. Clean the interior of the heat exchangers using a long flexible handle brush and a vacuum cleaner.

2. Clean the burners by gently striking the orifice end on a block of wood. This should remove any dirt or lint buildup in the tube.
    a. Replace the burners if they are extremely rusted, crushed, or if the burner ports have collapsed.

3. Use a brush and a stiff wire to clean the crosslighter.

4. Reassemble the components in reverse order (see Reassembly Instructions).

Secondary Heat Exchanger (Interior Cleaning) Disassembly

The following parts and assemblies must be removed before the heat exchanger can be cleaned (refer to Figure 3.11).

1. Loosen the clamps on the inlet and outlet connectors on the combustion air blower.

2. Remove the screws holding the combustion air blower to the blower deck.

3. Gently wiggle and pull the blower outward to disengage it from the inlet and outlet connectors.
    a. The blower can be pulled out so that it just clears the furnace casing on the left side without disconnecting any wiring. Support the blower so it does not hang by the wiring.

4. Remove the screws holding the electronic spark module to the blower deck.

5. Remove the screws attaching the secondary heat exchanger panel and remove the panel.

6. Remove the screws holding the Z bracket to the division panel and remove the bracket.
7. Loosen the screw on the secondary heat exchanger inlet coupling by reaching in on the right side.

> **Warning:** *Danger of bodily injury. Coil has sharp fins. Cover with rags and handle with care to avoid cuts.*

8. Remove the secondary heat exchanger by pulling straight out and reposition the spark module and wiring, etc., to get past them.

Cleaning. Use the following steps in cleaning the coil:

1. Use a one and five-eighths in. plastic cap plug or the palm of your hand to plug either the inlet or outlet port and fill the heat exchanger coil with approximately one and one-half quarts of hot water.
2. Plug the other port and shake the coil vigorously. Drain and flush with a hard stream of water from a garden hose. Repeat steps 1 and 2 if required.
3. Thoroughly wash the exterior. *Do not* use a hard stream of water on the exterior, because it will bend the coil fins.

Reassembly

Reassemble all the parts in reverse order as removed, with the following instructions.

Crosslighter. Install the burner into the slots of the crosslighter and press it into position. The back of the crosslighter must be seated firmly against the burner ports.

Burners. Insert the burners into the heat exchanger. The burners must be inserted into the slots at the back of the heat exchanger and be level.

Pilot Ignitor. Check and adjust the spark gap for the pilot ignitor (see Figure 3.13).

Flue Baffles. The flue baffles are to be installed into position as illustrated in Figure 3.14. The baffle must be located below the dimple in the heat exchanger and firm against the bottom of the flue outlet.

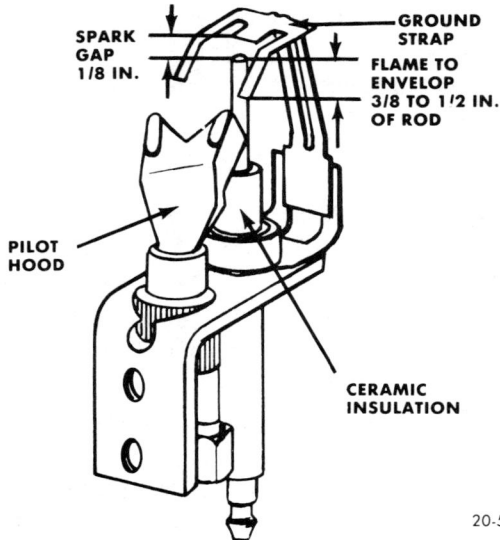

20-52-23

Figure 3.13 Pilot and ignitor assembly. (Courtesy of Heil-Quaker Corporation.)

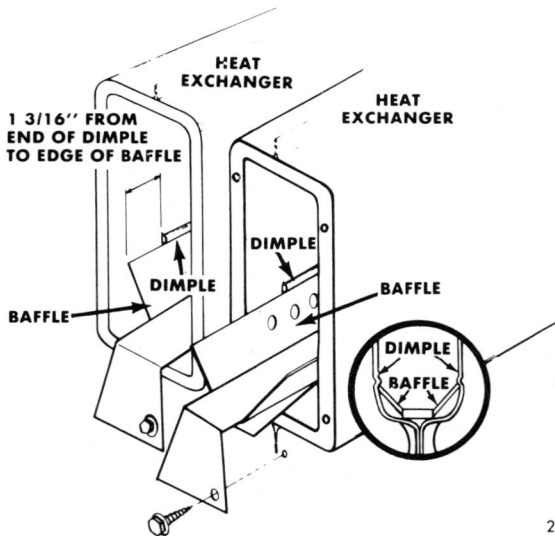

23-40-12

Figure 3.14 Installing flue baffles. (Courtesy of Heil-Quaker Corporation.)

Any insulation that is torn or defective must be repaired or replaced as necessary.

Replace all gaskets and parts that are broken or deteriorated.

Testing for Leaks.   After reassembly, turn the gas on and check all points for gas leaks using a soapy solution. All leaks must be repaired immediately. Perform an operational check of the furnace.

SPARK MODULE SEQUENCE OF OPERATION

The following is the sequence of operation for spark ignition of the pilot burner:

1. On a call for heat from the thermostat, 25 V are sent through the control and safety switches to the spark control module.
2. The spark module then sends electrical power to open the pilot portion of the gas valve and at the same time sends 15 000 V to the pilot ignitor to initiate a spark (see Figure 3.15).

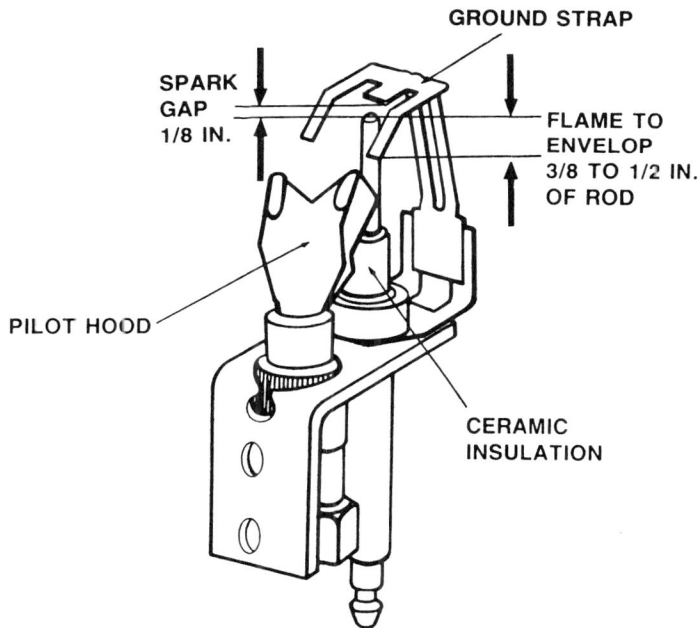

Figure 3.15 Pilot ignitor. (Courtesy of Heil-Quaker Corporation.)

3. When the pilot flame has been proven, a microampere signal is sent back to the control module so that the spark is terminated. On LP models, the pilot flame must be proven within 90 seconds, or the control module will lockout and must be reset by breaking the voltage to the control module.
4. When the module receives the microampere signal that a flame has been proven, the sparking stops and voltage is sent to the gas valve opening the main valve and lighting the main burner. As long as the pilot is burning, the microampere signal will keep the gas valve open and the burner on.
5. When the thermostat is satisfied, its contacts open, voltage is no longer

sent to the control module, and both the pilot and the main gas valve close, shutting down the furnace.

POWER VENT SYSTEM SINGLE SWITCH ASSEMBLY

The following is a description of the normal operation and checking procedures for these types of systems.

Normal Operation

As the circulating air fan starts turning, a positive (+) pressure starts to build up in the chamber of the power vent (see Figure 3.16). The sensor tube picks up the pressure where it is felt in the pressure side of the air switch.

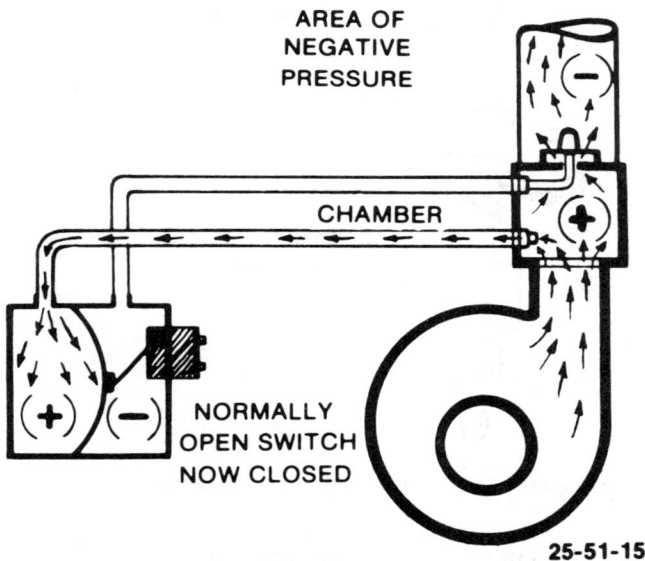

Figure 3.16 Power vent system single switch. (Courtesy of Heil-Quaker Corporation.)

The pressure moves the diaphragm, closing the normally open switch, completing the 24-VAC path to the D. S. I module.

As the positive (+) air pressure builds up in the chamber of the power vent, notice the venturi in the center of the chamber with Pitot tube or J tube extending into the center of the venturi. As the positive (+) or pressurized air passes the end of the tube, a small area of negative (−) pressure is formed above the tube. The negative (−) pressure is felt in the blockage side of the air switch and helps hold the switch closed.

Checking Flue Vent Pressures

Use the following steps to check the flue vent pressures:

1. Using a U tube manometer or a Magnehelic gauge, remove the tube from the pressure switch and attach the gauge (see Figure 3.17).

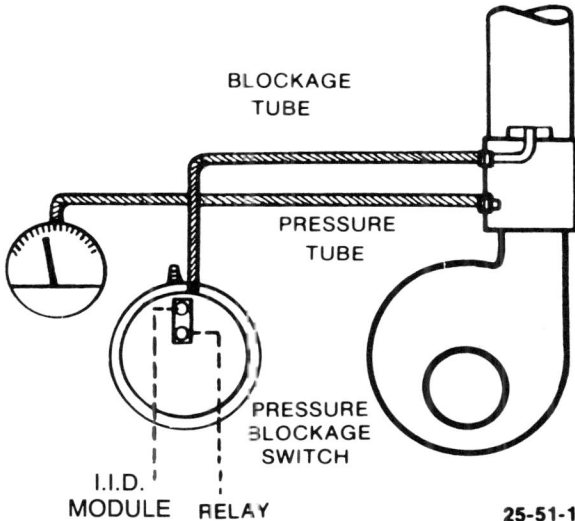

BLOCKAGE
TUBE

PRESSURE
TUBE

PRESSURE
BLOCKAGE
SWITCH

I.I.D.
MODULE   RELAY   **25-51-19**

Figure 3.17 Checking flue vent pressures. (Courtesy of Heil-Quaker Corporation.)

2. With the vent fan operating, the normal pressure should be above 0.62 in. wc (depending upon the vent temperatures). Once the normally open switch is closed it should not open unless the pressure drops below 0.45 to 0.55 in. wc.

3. If the pressure reads 0.62 in. wc and the switch does not close, remove the tubing from the blockage side of the switch. If the switch still does not close, replace the switch assembly.

4. If the pressure is below the 0.62 in. wc normal operating pressure, check for soot in the heat exchanger and on the blower wheel or broken sensor tube. Clean and correct the condition causing the soot to build up. Replace the sensor tube if necessary.

Pitot (J) Tube Adjustment and Blockage Pressures

Use the following procedures to check the blockage pressures and tube adjustment:

1. The J tube must be centered in the orifice (see Figure 3.18). The tube must extend one-fourth in. above the orifice plate opening.

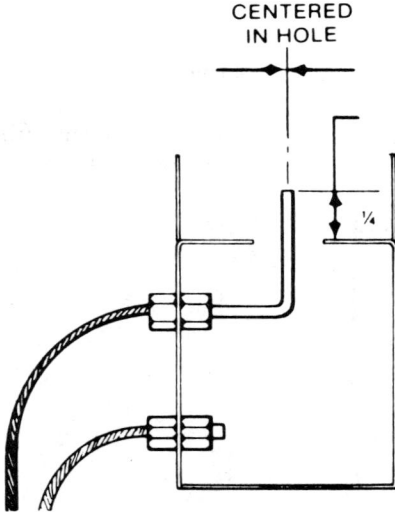

Figure 3.18 Pilot J tube adjustment and blockage pressures. (Courtesy of Heil-Quaker Corporation.)

2. The tube must be free of burrs, both inside and out.

   Note: Caution should be taken to prevent shavings from entering the switches when deburring.

3. Using a U tube manometer, or Magnehelic gauge, check the flue pressures.

Under normal operating conditions, a negative ($-$) pressure of at least 1 in. wc should be read at the switch. If you cannot obtain this negative pressure, check the condition of the vent (too many elbows, flue too long, partially obstructed flue, etc.). In the case of flue blockage, the switch will open when the pressure exceeds 0.12 in. wc.

Table 3.6 gives the pressure switch data for certain model Heil furnaces.

TABLE 3.6   PRESSURE SWITCH DATA (COURTESY OF HEIL-QUAKER CORPORATION)

Switch number	Used on	Set point	Color code
1000743	NDG/LK050	4.12	Yellow
1000744	NDG/LK075	3.35	Orange
1000745	NDG/LK100	2.36	Red
1000746	NDG/LK125	1.91	Blue
1000747	NUG/LK050	4.175	Green
1000748	NUG/LK075	3.2	Grey
1000749	NUG/LK125	1.95	Purple
1001181	NUG/LK100	2.3	White
611921	Previous NUG/LK 1985 "K" All Models	2.5	

HEIL GAS FURNACE UNIT WIRING DIAGRAMS

GAS UPFLOW I.I.D.

ALL MODELS

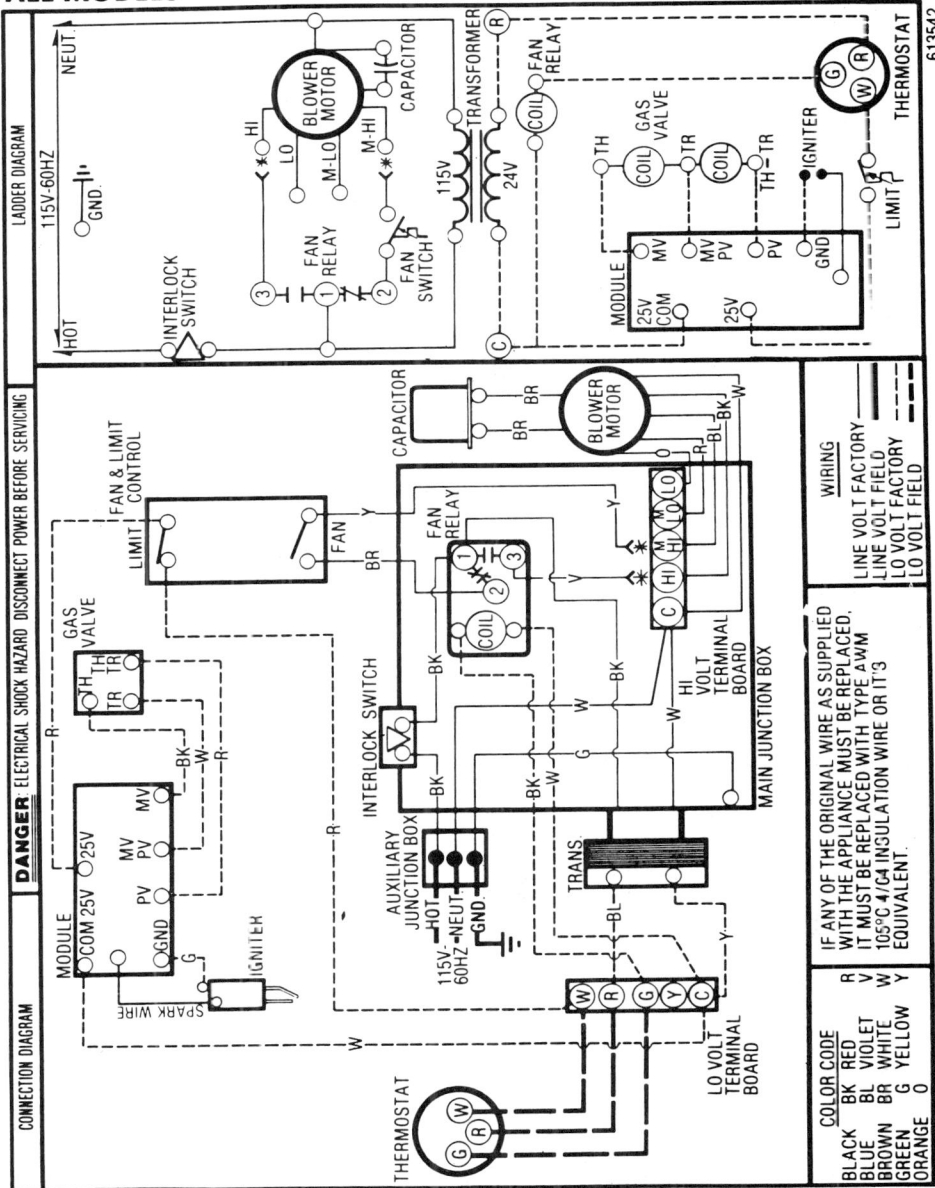

WIRING DIAGRAM—S.P.D.T. RELAY

WIRING DIAGRAM—D.P.D.T. RELAY

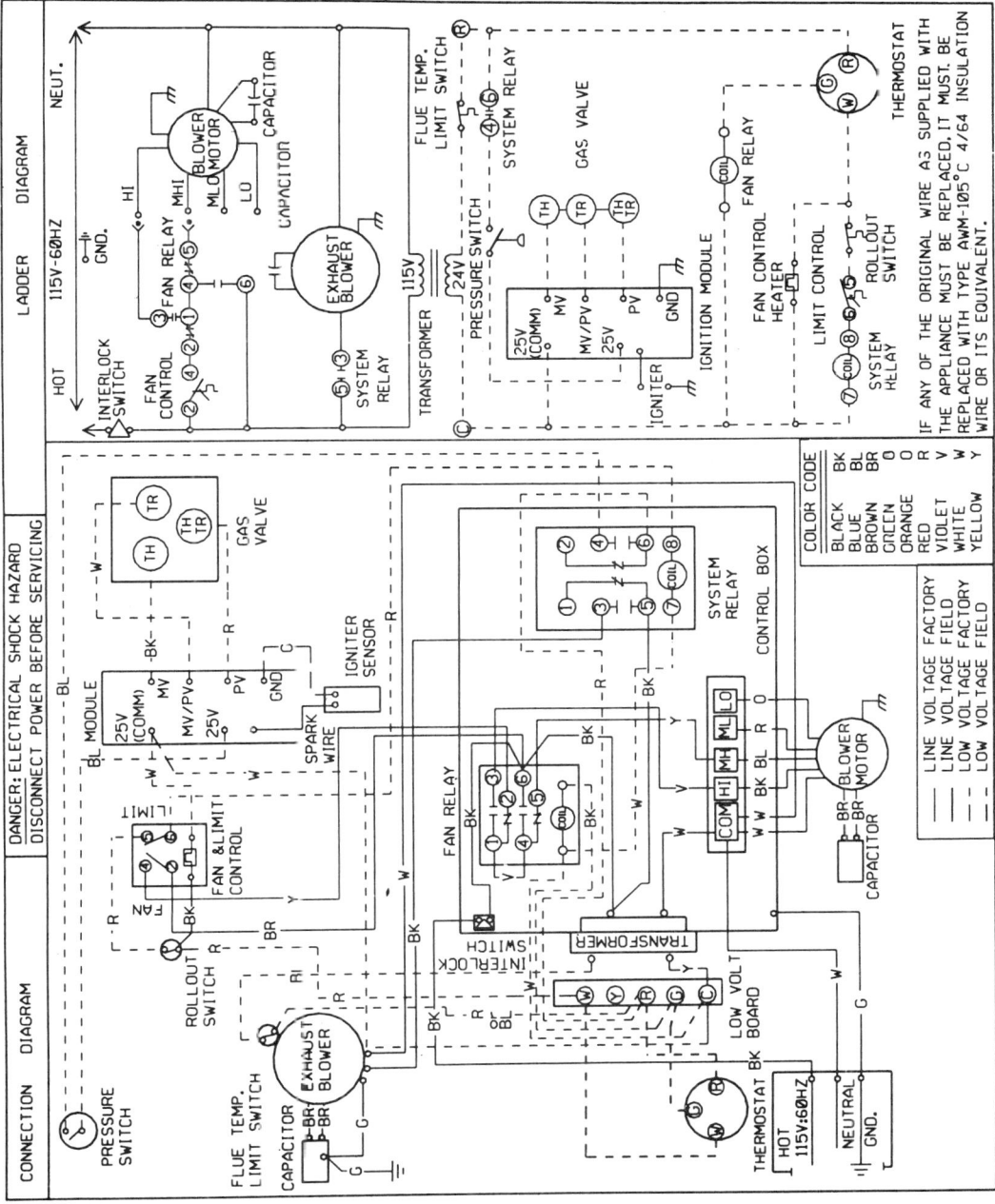

SERVICE ANALYZER CHART

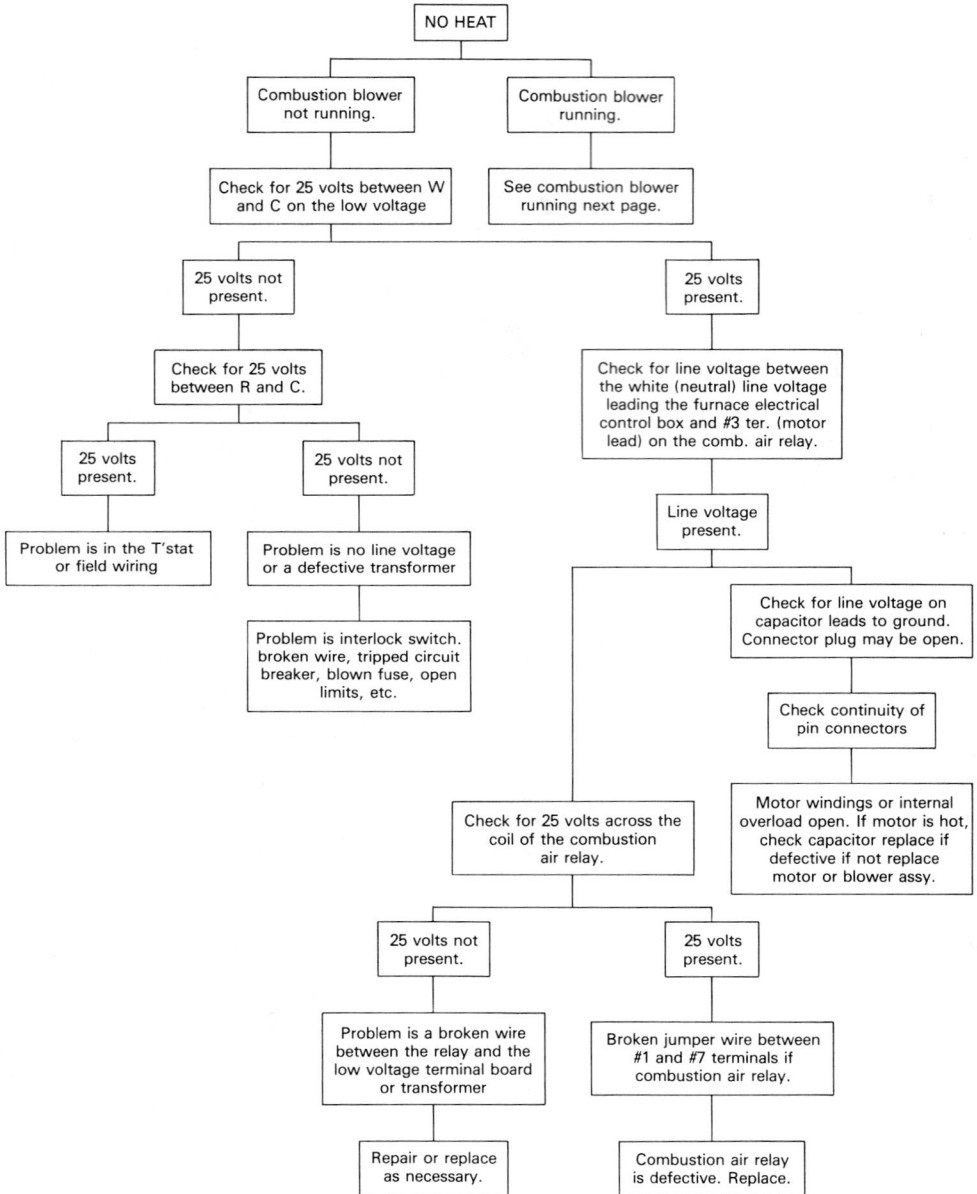

```
                              ┌─────────────┐
                              │  NO HEAT    │
                              └─────────────┘
                 ┌──────────────────────┴──────────────────────┐
       ┌───────────────────┐                      ┌───────────────────┐
       │ Combustion blower │                      │ Combustion blower │
       │   not running.    │                      │     running.      │
       └───────────────────┘                      └───────────────────┘
                 │                                           │
   ┌───────────────────────────┐            ┌───────────────────────────┐
   │ Check for 25 volts between W│            │ See combustion blower     │
   │ and C on the low voltage   │            │ running next page.        │
   └───────────────────────────┘            └───────────────────────────┘
```

```
   ┌─────────────────┐                          ┌─────────────────┐
   │ 25 volts not    │                          │ 25 volts        │
   │ present.        │                          │ present.        │
   └─────────────────┘                          └─────────────────┘
```

Check for 25 volts between R and C.

25 volts present.

25 volts not present.

Check for line voltage between the white (neutral) line voltage leading the furnace electrical control box and #3 ter. (motor lead) on the comb. air relay.

Line voltage present.

Problem is in the T'stat or field wiring

Problem is no line voltage or a defective transformer

Problem is interlock switch. broken wire, tripped circuit breaker, blown fuse, open limits, etc.

Check for line voltage on capacitor leads to ground. Connector plug may be open.

Check continuity of pin connectors

Motor windings or internal overload open. If motor is hot, check capacitor replace if defective if not replace motor or blower assy.

Check for 25 volts across the coil of the combustion air relay.

25 volts not present.

25 volts present.

Problem is a broken wire between the relay and the low voltage terminal board or transformer

Broken jumper wire between #1 and #7 terminals if combustion air relay.

Repair or replace as necessary.

Combustion air relay is defective. Replace.

SERVICE ANALYZER CHART

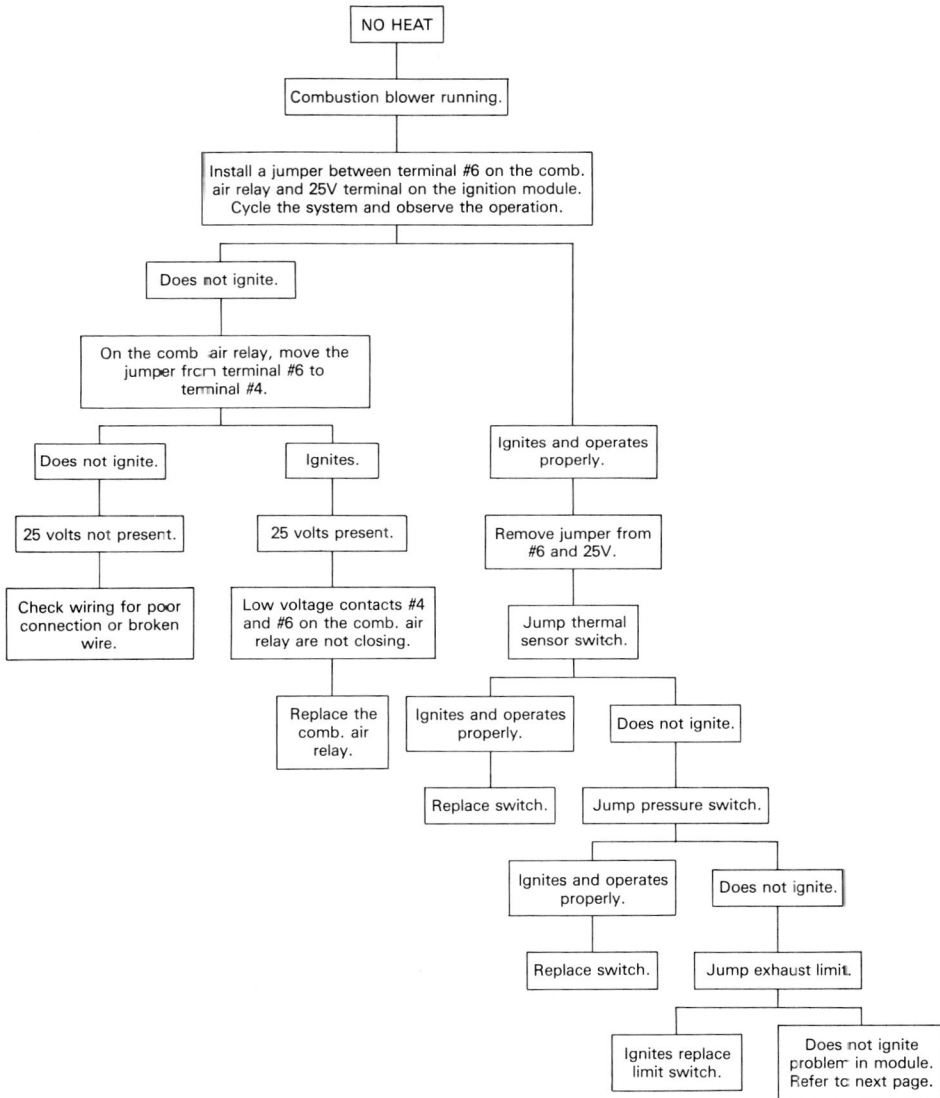

NO HEAT

Combustion blower running.

Install a jumper between terminal #6 on the comb. air relay and 25V terminal on the ignition module. Cycle the system and observe the operation.

Does not ignite.

On the comb. air relay, move the jumper from terminal #6 to terminal #4.

Does not ignite.

Ignites.

Ignites and operates properly.

25 volts not present.

25 volts present.

Remove jumper from #6 and 25V.

Check wiring for poor connection or broken wire.

Low voltage contacts #4 and #6 on the comb. air relay are not closing.

Jump thermal sensor switch.

Replace the comb. air relay.

Ignites and operates properly.

Does not ignite.

Replace switch.

Jump pressure switch.

Ignites and operates properly.

Does not ignite.

Replace switch.

Jump exhaust limit.

Ignites replace limit switch.

Does not ignite problem in module. Refer to next page.

4

Lennox Pulse™
GSR14
Gas Furnace

INTRODUCTION

The GSR14 unit is a condensing gas furnace utilizing the pulse combustion process. It is unique in that it may be mounted in either the down-flow or the horizontal flow position. The combustion process is identical to that in the G14 series Pulse furnaces. Initially combustion takes place in an enclosed chamber. Then, as combustion products pass through the heat exchanger system into a coil, the latent heat of combustion is extracted by condensing water from the exhaust gas (see Figure 4.1).

The unit uses a redundant gas valve to assure safety shutoff, as required by the American Gas Association (AGA).

An electronic direct spark ignition system is used to initiate the combustion process. A flame rectification sensor verifies ignition with a protection circuit that permits five trials for ignition before "locking out" the gas valve and control circuit. The sensor also verifies loss of combustion during a heating cycle, closing the gas valve, and locking out the system. Obstructions to the air intake or exhaust outlet will also cause the unit to shut down immediately.

A small blower is used to purge the combustion chamber before and after each heating cycle to provide proper air mixture for start-up.

These units are manufactured for natural gas; LP kits are available for field modification.

110

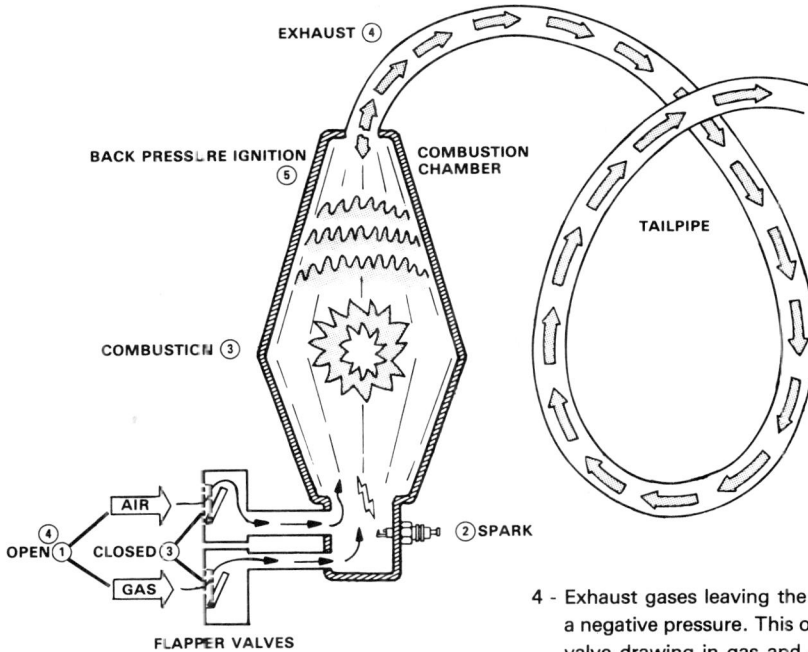

Figure 4.1 Pulse combustion process. (Courtesy of Lennox Industries, Inc.)

1 - Gas and air enter and mix in combustion chamber.
2 - To start the cycle a spark is used to ignite the gas and air mixture (This is one 'pulse').
3 - Positive pressure from combustion closes flapper valves and forces exhaust gases down a tailpipe.
4 - Exhaust gases leaving the chamber create a negative pressure. This opens the flapper valve drawing in gas and air.
5 - At the same instant part of the pulse is reflected back from the tailpipe causing the new gas and air mixture to ignite. No spark is needed (This is another 'pulse').
6 - Steps 4 and 5 repeat 60 to 70 times per second forming consecutive 'pulses' of 1/4 to 1/2 Btu each.

UNIT COMPONENTS

The following is a description of the components that make up the Lennox GSR14 furnace (see Figure 4.2).

Makeup Box

The makeup box is located behind the blower access panel (see Figure 4.3). The components located in the makeup box are as follows:

1. Low-voltage terminal strip with thermostat markings.
2. A 30-VA transformer, 120-VAC primary/24-VAC secondary. The trans-

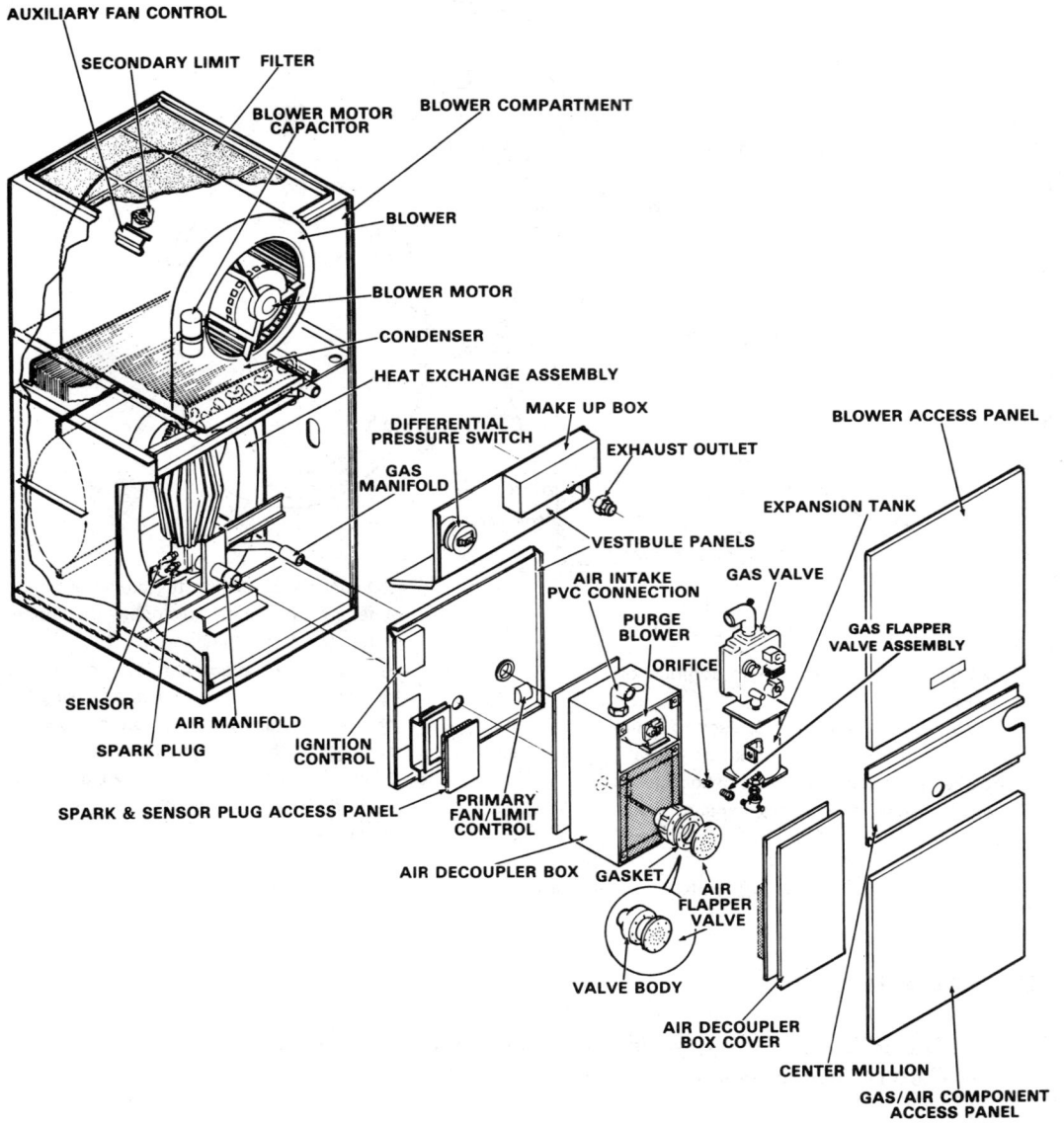

Figure 4.2 Parts arrangement. (Courtesy of Lennox Industries, Inc.)

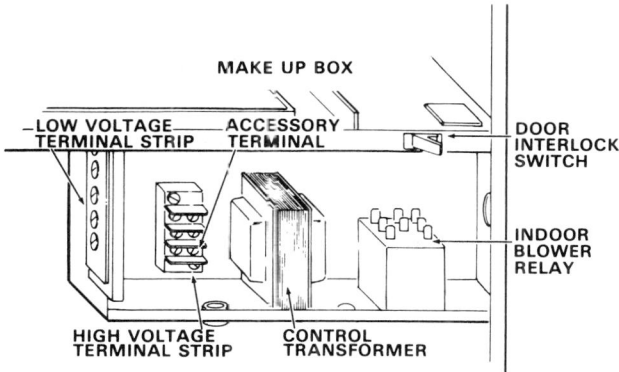

Figure 4.3 Makeup box. (Courtesy of Lennox Industries, Inc.)

former located inside the makeup box provides power to the low-voltage section of the unit. The transformers on all models are rated at 30 VA with a 120-VAC primary and a 24-VAC secondary. On CGA (Canadian Gas Association) models the transformer includes a 2.0-A fuse on the "load" side of the secondary.

3. A double-pole, double-throw indoor blower relay. An indoor blower double-pole, double-throw relay is located inside the makeup box to provide power to the blower motor. The relay contacts also control the 120-VAC accessory terminal which is located on the high-voltage terminal strip inside the makeup box.
4. A power supply and accessory terminal strip.

Fan/Limit Control

A fan/limit control with a sure start heater is used to control the blower motor operation. It is located in the lower end of the unit next to the air intake chamber. The fan control heater is energized with the gas valve and its contacts close after 30 to 45 seconds to bring on the blower (see Figure 4.4).

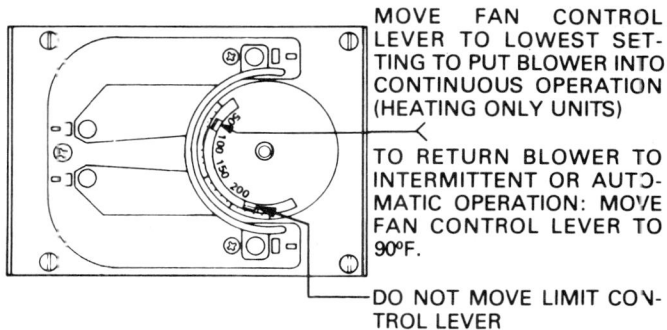

MOVE FAN CONTROL LEVER TO LOWEST SETTING TO PUT BLOWER INTO CONTINUOUS OPERATION (HEATING ONLY UNITS)

TO RETURN BLOWER TO INTERMITTENT OR AUTOMATIC OPERATION: MOVE FAN CONTROL LEVER TO 90°F.

DO NOT MOVE LIMIT CONTROL LEVER

Figure 4.4 Fan/limit control location. (Courtesy of Lennox Industries, Inc.)

The fan OFF setting is factory adjusted to 90 °F. It should not be necessary to change this setting. Fan OFF settings above 90 °F will cause the blower to recycle frequently (after a heating cycle) due to the additional heat left over in the heat exchanger assembly.

Do not change the limit factory setting. This is fixed in position for a maximum discharge air temperature of 175 °F. This is a safety shutdown function. The limit will automatically reset when the temperature inside the heat exchanger housing drops below the temperature listed above.

Auxiliary Fan Control

An auxiliary fan control is located inside the blower housing next to the secondary limit. Its purpose is to protect the secondary limit from "tripping" by turning on the blower until the blower compartment temperature is reduced. This is a safety cool-down function of the unit and should serve to prevent service calls to reset the secondary limit (see Figure 4.5).

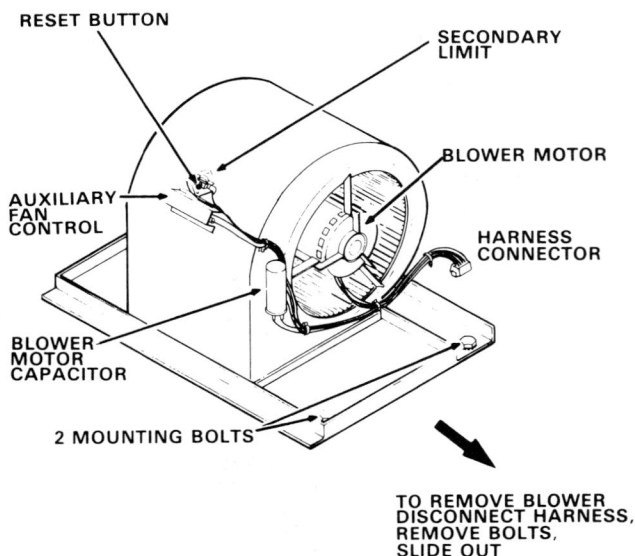

Figure 4.5 Auxiliary fan control. (Courtesy of Lennox Industries, Inc.)

> **Warning:** *Shock hazard. The auxiliary fan control is connected to line voltage. It is housed under a barrier for safety. Before servicing the control, be sure to disconnect the electrical power to the unit.*

> **Warning:** *The barrier that shrouds the auxiliary fan control must be in place any time the unit is operating. Do not operate the unit without the barrier in place.*

Secondary Limit

A secondary limit control is located on the blower housing (see Figure 4.5). Its purpose is to stop the unit in the event the blower does not operate properly. If the blower should fail to operate or if either the return or supply airways become partially blocked, the blower housing will become too warm due to heat flow and cause the secondary limit to "trip." The secondary limit can be reset manually by pressing the button on the back of the limit control. Allow adequate time for the limit to cool before attempting to reset. This is a safety shutdown function of the unit.

The secondary limit and the auxiliary fan control work together to reduce excessive temperature in the upper (blower) end of the unit. First, as temperature rises in the blower compartment and nears 140 °F, the auxiliary fan control actuates the blower in an attempt to reduce the temperature. Should the blower be unable to reduce the temperature in the blower housing and the temperature continue to rise, the secondary limit will activate and "shut down" the unit.

Ignition Control

Lennox GSR14 furnaces are equipped with an electronic ignition control. The following is a description of the ignition controls used on these furnaces:

1. An electronic direct spark ignition with flame rectification sensing is used on all GSR14 units. The units may have ignition controls manufactured by Lennox, Gas Energy, or Watsco (formerly Prestolite).

   > **Danger:** *Shock hazard. Spark-related components contain high voltage. Disconnect the electrical power before servicing. The ignition control is not field repairable. Unsafe operation may result.*

2. The Lennox, Watsco, and Gas Energy ignition controls are interchangeable between units. Any control can connect to any unit using the harness plug (JP-1). The Gas Energy control uses an interconnecting harness to connect to the JP-1. Spark and sensor wires for each control are separate and different from the JP-1 (see Note later and Figures 4.6–4.8).

Note: Replacement ignition controls may be made by a different manufacturer than the control being removed. The spark and sensor connections *may* be different. Ignition control replacement kits come supplied with new spark and sensor wires which fit the replacement control. Replacement wires with connectors may also be ordered separately. *Do not attempt to cut or splice connectors on old wires.* Refer to Lennox repair parts for detailed ordering information and part numbers.

3. The Lennox GC-1 control is illustrated in Figure 4.6. The unit wiring harness plugs directly into the jack at the corner of the control. Each of the six jack terminals are identified by number and function. The spark plug wire connection is made to the spark plug-type connector on the control as shown. The sensor wire connection is made to a quick connect terminal on the control as shown.

   Important: If the Lennox GC-1 control is used, *a ceramic resistor spark plug must also be used.*

4. The Watsco control is illustrated in Figure 4.7. The unit wiring harness plugs directly into the jack at the lower right corner of the control. Each

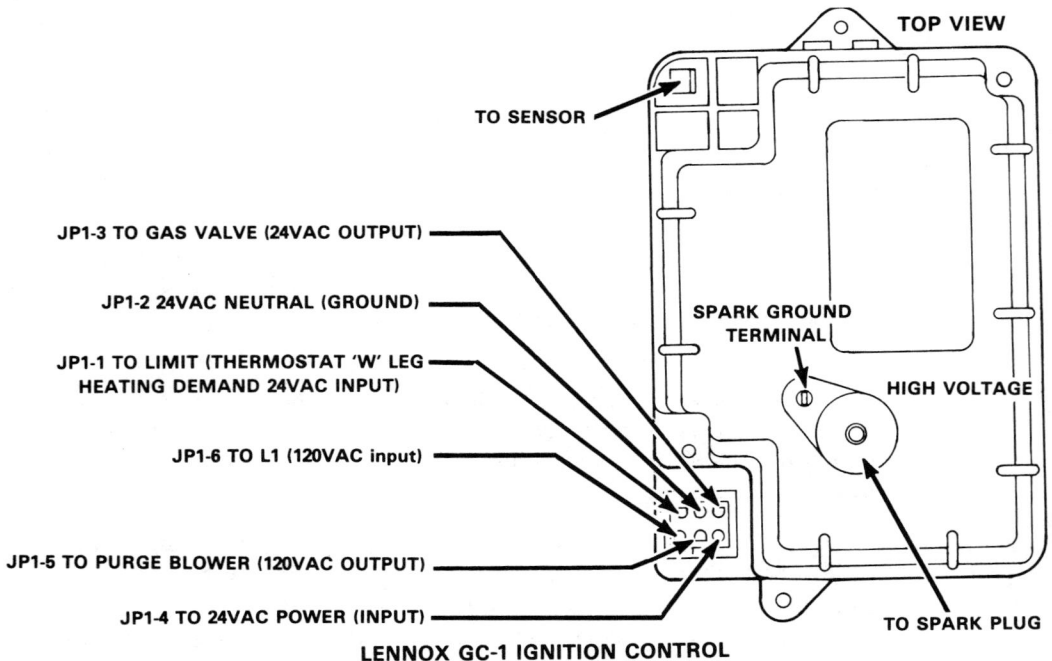

TOP VIEW

TO SENSOR ►

JP1-3 TO GAS VALVE (24VAC OUTPUT) ──

JP1-2 24VAC NEUTRAL (GROUND) ──

SPARK GROUND TERMINAL

JP1-1 TO LIMIT (THERMOSTAT 'W' LEG HEATING DEMAND 24VAC INPUT) ──

HIGH VOLTAGE

JP1-6 TO L1 (120VAC input) ──

JP1-5 TO PURGE BLOWER (120VAC OUTPUT) ──

JP1-4 TO 24VAC POWER (INPUT) ──

TO SPARK PLUG

LENNOX GC-1 IGNITION CONTROL

Figure 4.6 Lennox GC-1 ignition control. (Courtesy of Lennox Industries, Inc.)

TOP VIEW

JP1-3 TO GAS VALVE (24VAC OUTPUT)
JP1-2 TO 24VAC NEUTRAL (GROUND)
JP1-1 TO LIMIT (THERMOSTAT ' W" LEG
HEATING DEMAND 24VAC INPUT)

JP1-4 TO 24VAC POWER (INPUT)
JP1-5 TO PURGE BLOWER (120VAC OUTPUT)
JP1-6 TO L1 (120VAC INPUT)

HIGH VOLTAGE
SENSE
TO SPARK PLUG
TO SENSOR

WATSCO IGNITION CONTROL

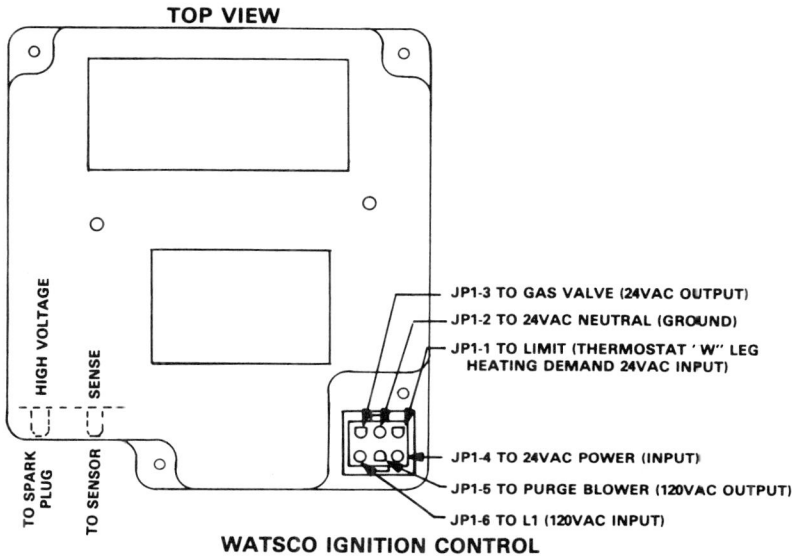

Figure 4.7 Watsco ignition control. (Courtesy of Lennox Industries, Inc.)

of the six jack terminals are identified by number and function. The spark and sensor wire connections use separate quick connect terminals on the control as shown.

5. The Gas Energy control is illustrated in Figure 4.8. This control has an interconnecting harness used to connect it to the unit harness wiring plug. The terminals and connections are identified by function and number. The spark connection is made to the spark plug-type connector on the control as shown.

6. The primary control provides five main functions: prepurge, control of the gas valve, ignition, flame sensing, and postpurge. The ignition attempt sequence of the control provides five trials for ignition before locking out. The unit will usually ignite on the first attempt. See Figure 4.9 for a normal ignition sequence with nominal timings for simplicity.

7. Proper gas/air mixture is required for ignition on the first attempt. If there is a slight deviation, within the tolerance of the unit, a second or third trial may be necessary for ignition. The control will lock out the system if ignition is not obtained within five trials. Reset after lockout requires only breaking and remaking the thermostat demand. See Figure 4.10 for the ignition attempt sequence with trials (nominal timings given for simplicity). Loss of combustion during the heating cycle is sensed through an absence of a flame signal causing the control to lockout after five ignition trials.

TOP VIEW

TO SENSOR

T6 TO PURGE BLOWER (120VAC OUTPUT)

T5 TO L1 (120VAC INPUT)

JP1

BROWN1		4BLUE
YELLOW2		5RED
ORANGE3		6BLACK

RED

BLACK

ORANGE

BLUE

BROWN

YELLOW

1
2
3

INTERCONNECTING
WIRING HARNESS TO MATE
CONTROL TO UNIT HARNESS

JP1

HIGH VOLTAGE

T4 TO GAS VALVE (24VAC OUTPUT)

T1 TO 24VAC POWER (INPUT)

T3 TO LIMIT (THERMOSTAT 'W' LEG HEATING DEMAND 24VAC INPUT)

T2 24VAC NEUTRAL (GROUND)

TO SPARK PLUG

FUSE
(125V,3A,slow blow type)

GROUND STRAP

SPARK PLUG WIRE →

GAS ENERGY IGNITION CONTROL

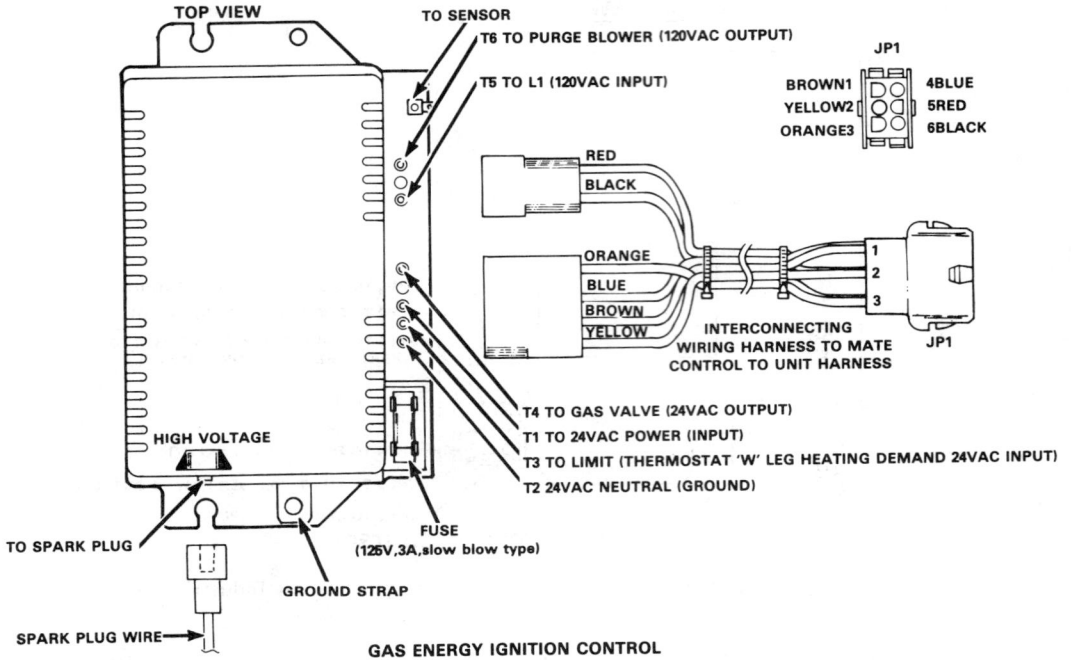

Figure 4.8 Gas Energy ignition control. (Courtesy of Lennox Industries, Inc.)

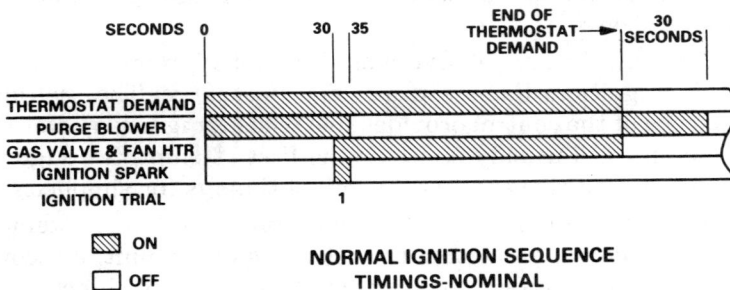

SECONDS   0   30   35   END OF   30
                                      THERMOSTAT → SECONDS
                                      DEMAND

THERMOSTAT DEMAND
PURGE BLOWER
GAS VALVE & FAN HTR
IGNITION SPARK

IGNITION TRIAL   1

▨ ON
☐ OFF

NORMAL IGNITION SEQUENCE
TIMINGS-NOMINAL

1 - Thermostat demand for heat.
2 - Purge blower is energized.
3 - At 30 seconds gas valve, fan heater and ignition spark are energized for 5 seconds.
4 - When ignition occurs (sensed by flame rectification), the

spark and purge blower are de-energized.
5 - At end of heating demand, gas valve and fan heater are de-energized and purge blower is started.
6 - Post purge continues for 30 seconds after heating cycle, then is de-energized.

Figure 4.9 Normal ignition sequence timings, nominal. (Courtesy of Lennox Industries, Inc.)

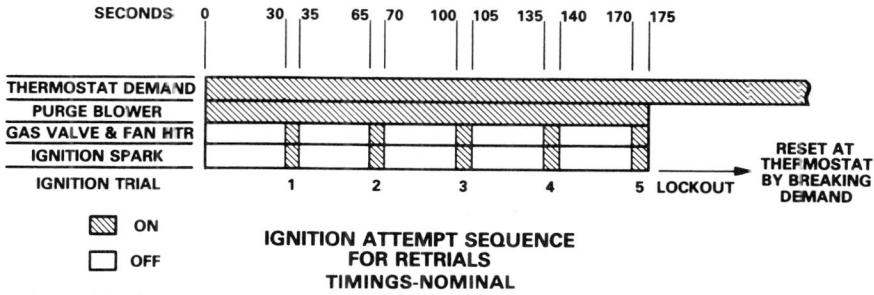

Figure 4.10 Ignition attempt sequence for trials timings, nominal. (Courtesy of Lennox Industries, Inc.)

8. The specific timings for the Lennox GC-1, Gas Energy, and Watsco ignition controls vary, but do not affect operation of the unit. All will make five trials for ignition before lockout. The specific timings for each are given in Figure 4.11.

   The Watsco control runs through a postpurge cycle each time power is interrupted to the unit, for example, when the main switch is turned off and on again, following intermittent power failures, or when replacing the blower door, thus energizing the interlock switch. This is a normal characteristic unique to the Watsco control.

9. The Gas Energy control uses an externally mounted integral fuse to protect the circuitry from accidental shorting while servicing the unit. The fuse, mounted on the lower right corner of the control, is rated 125-V, 3-A slow blow type. Refer to the Lennox Repair Parts for fuse replacement.

Gas Valve and Expansion Tank

The following is a description of these components (see Figure 4.12):

1. All units use a gas valve with a 1-second or less opening time from zero to maximum manifold pressure. The valve is internally redundant to assure safety shutoff. There are six adjacent terminals that are grouped in pairs on top of the valve. The terminals are used to energize the internal redundant solenoids. The solenoids are wired in parallel. A manual shutoff knob is provided on the valve. Note that terminals 1 and 3 are load and are jumpered together. Terminal 2 is common.

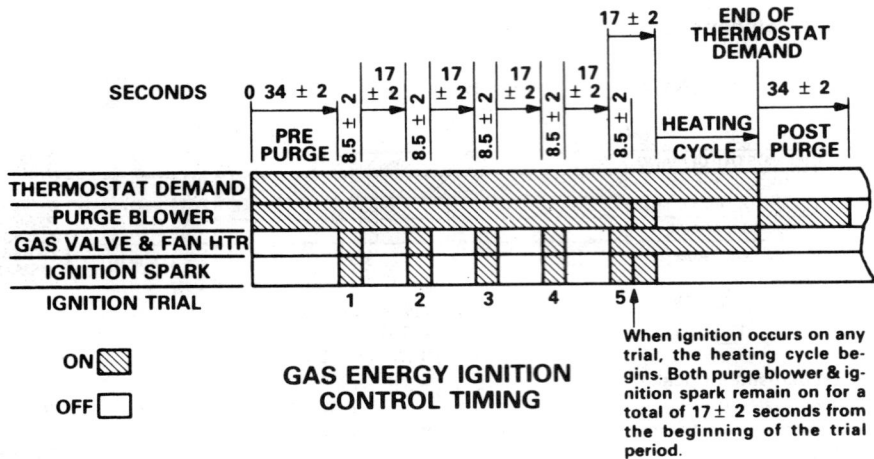

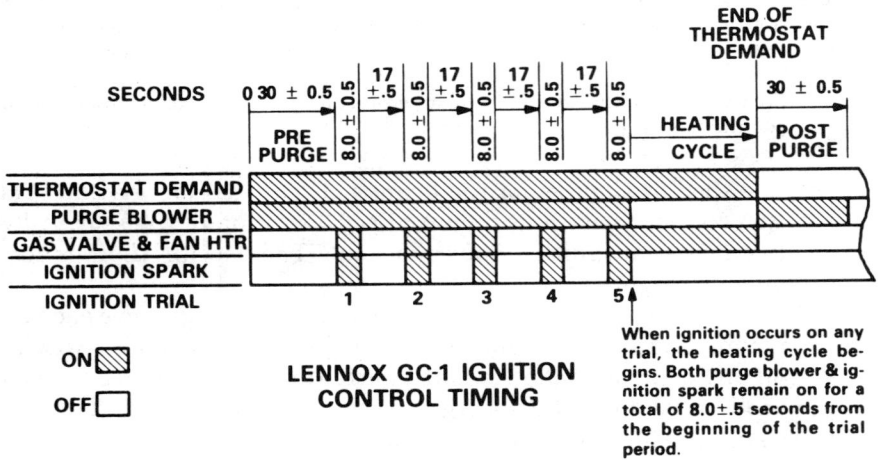

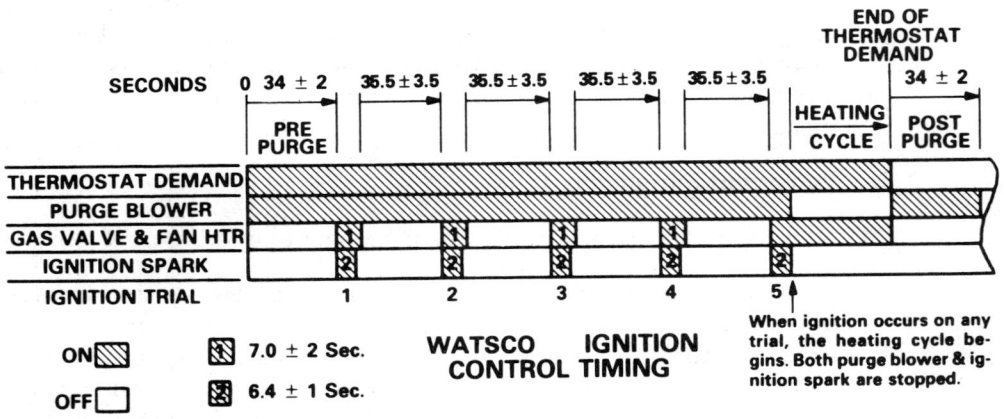

Figure 4.11 Gas Energy, Lennox GC-1, and Watsco ignition control timing. (Courtesy of Lennox Industries, Inc.)

120

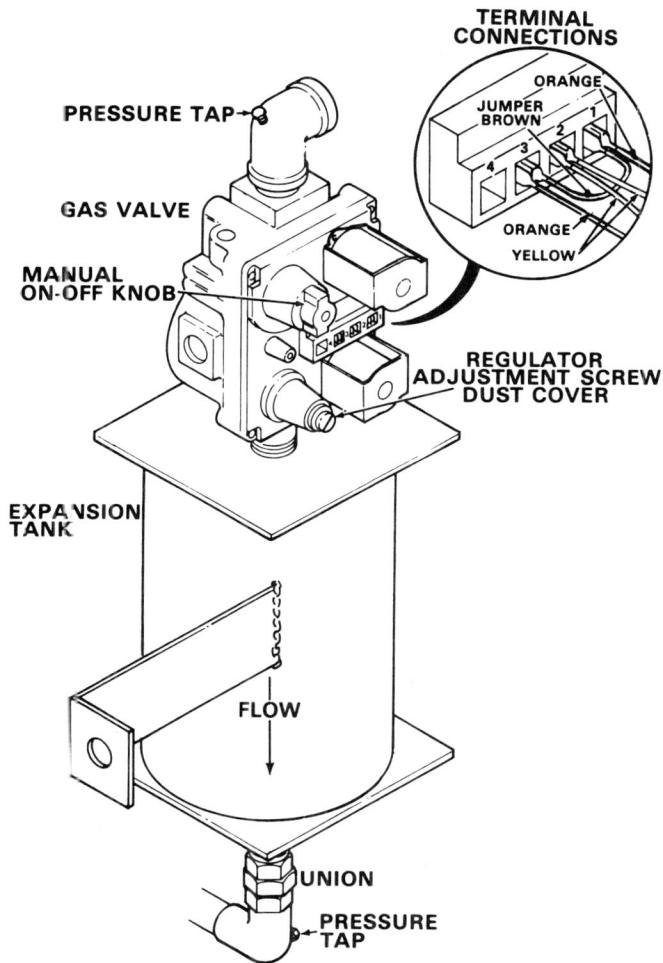

Figure 4.12 Gas valve and expansion tank. (Courtesy of Lennox Industries, Inc.)

2. The expansion tank located downstream of the gas valve absorbs any back pressure created during combustion to prevent damage to the gas valve diaphragm.

Differential Pressure Switch

A differential pressure switch is mounted next to the makeup box (see parts arrangement, Figure 4.2). It is connected to the air intake and exhaust outlet by separate lengths of flexible plastic tubing. Note that each flexible hose connects to a barbed adaptor at the differential pressure switch (see Figure 4.13). Each adaptor has a built in orifice of 0.018 in. i.d.

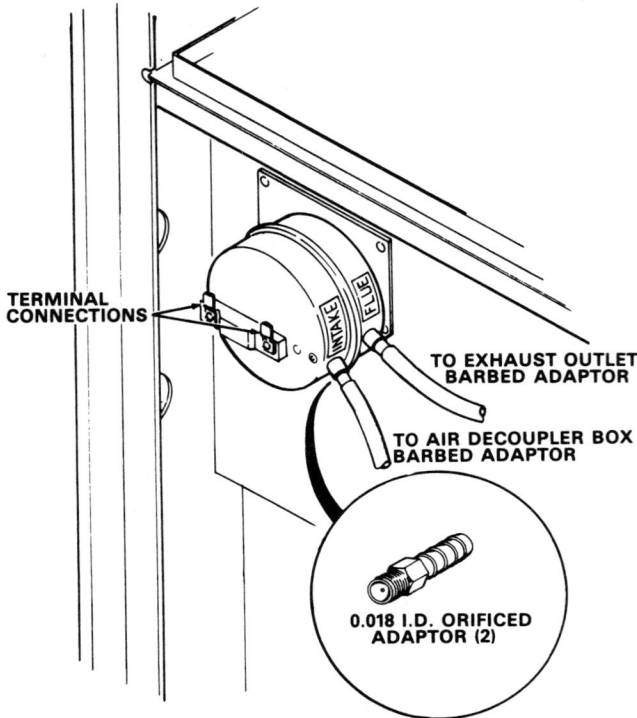

TERMINAL CONNECTIONS

INTAKE   FLUE

TO EXHAUST OUTLET BARBED ADAPTOR

TO AIR DECOUPLER BOX BARBED ADAPTOR

0.018 I.D. ORIFICED ADAPTOR (2)

NOTE: ADAPTORS POINTED DOWNWARD FOR HOSE DRAINAGE

Figure 4.13 Differential pressure switch. (Courtesy of Lennox Industries, Inc.)

> **Caution:** Each orifice adaptor is critical to switch operation. The orifice reduces the extreme positive and negative pressure peaks and must be used to prevent erratic switch operation. *Do not replace with nonorificed adapters.* Replacement adapters are available through Lennox Repair Parts.

The switch is normally closed and remains closed under normal operating conditions. Any obstruction or close off of the air intake or exhaust outlet (a differential pressure condition) causes the switch to open. When the switch opens it breaks the heat demand circuit to shut down the unit. This is a safety function. The switch automatically resets when the restriction is removed from the air intake or exhaust.

The switch is factory preset to open at 3.0 ± 0.25 in. wc total differential pressure (the difference in pressure between the intake side and the exhaust side) and cannot be adjusted. Some GSR14-100's may be equipped with switches preset to open at 4.5 in. wc. Note that the switch is positioned with the orifice adapters pointing downward so condensate can drain from the switch.

Gas Intake Flapper Valve

The following is a description of the gas intake flapper valve (see Figure 4.14):

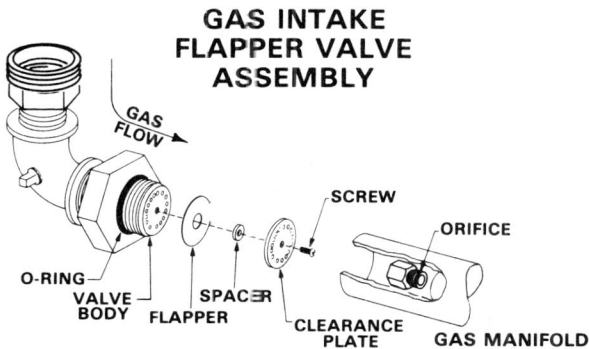

Figure 4.14 Gas intake flapper valve assembly. (Courtesy of Lennox Industries, Inc.)

1. A union at the bottom of the expansion tank (to the left of the expansion tank if mounted horizontally) provides for removal of the gas flapper valve assembly and access to the orifice.

   The flapper floats freely over the spacer and is opened against the back plate by incoming gas pressure. Back pressure from each combustion pulse forces the flapper against the valve body, closing off the gas supply.

   Refer to the troubleshooting section of this chapter for specific information about flapper valve inspection and conditions requiring replacement.

2. Each GSR14 unit uses only one orifice located downstream of the flapper valve. The orifice is sized specifically for each unit. High-altitude derating is not normally required. If special conditions exist where derating is required, contact the Lennox headquarters in your area.

   Note: Standard atmospheric burner orifices or orifice blanks cannot be used as replacements in GSR14 units.

   Note: Most orifices designed for G14 pulse furnaces cannot be used as replacements in GSR14 units. For proper orifice sizing please refer to Lennox Repair Parts.

Note: Proper orifice sizing is dependent on many variables. Each Pulse furnace is shipped with the proper orifice. For proper orifice sizing please refer to Lennox Repair Parts.

Air Intake and Purge Blower

The following is a description of the air intake and purge blower (refer to Figure 4.2):

1. The air intake chamber houses the purge blower and air intake flapper assemblies. Air enters through the air intake pipe (center of mullion) and passes through the purge blower and through the flapper valve to the combustion chamber.
2. The purge blower has a 120-V motor and is permanently lubricated. It is powered only during pre- and postpurge operation. During the Pulse combustion process the purge blower is not powered, but air is drawn through the purge blower by a negative pressure.

Air Intake Flapper Valve

The following is a description of the air intake flapper valve (see Figure 4.2):

1. The air intake flapper valve is similar to the gas flapper valve in operation. A flapper floats freely over a spacer between two plates. In actual operation, initially, the flapper is forced against the back plate by the purge blower, allowing air to enter the combustion chamber. Next, back pressure from the combustion forces the flapper against the cover plate, closing off the air supply. Finally, as a negative pressure is created in the combustion chamber, the flapper is drawn to the back plate and air enters. The back pressure and negative pressures control the flapper valve with each pulse once ignition has occurred.

Refer to the troubleshooting section of this chapter for specific information about flapper valve inspection and conditions requiring replacement.

Combustion Chamber and Heat Exchange Assembly

The following is a description of the combustion chamber and heat exchanger assembly (see Figure 4.15):

1. The combustion chamber has gas and air intake manifolds. The gas intake is on the right and the air intake is front-center (the air intake is above the gas intake on horizontally mounted units). The exhaust gas leaves through the tail pipe at the top (blower end) of the chamber.

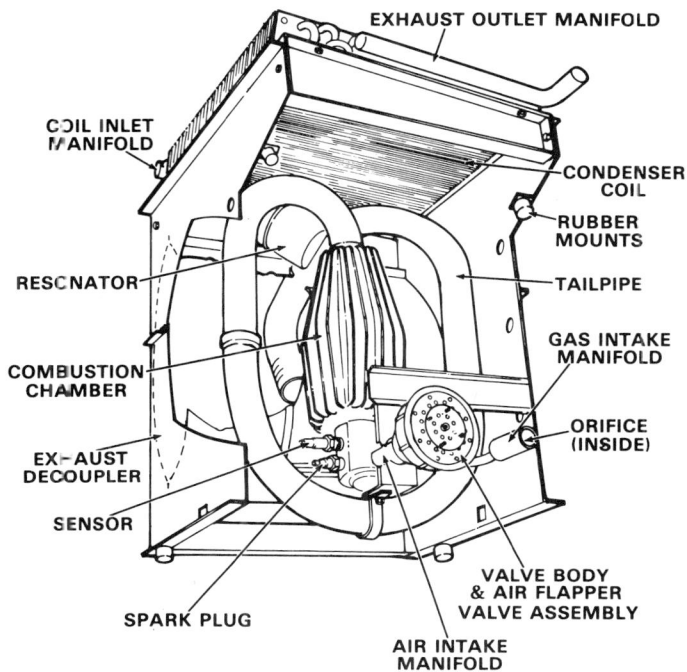

Figure 4.15 Combustion chamber and heat exchanger assembly. (Courtesy of Lennox Industries, Inc.)

2. The tail pipe connects the combustion chamber to the exhaust gas decoupler. The tail pipe and decoupler create the proper amount of back pressure for combustion to continue and are major heat exchange components. The resonator provides attenuation for acoustic frequencies.

3. The exhaust decoupler is manifolded to the condenser coil. The condenser coil is where the latent heat of combustion is taken from the exhaust gas. When this is done, condensate (moisture) is produced. The circuiting of the coil allows for the proper drainage of the condensate to the exhaust line. For this reason, it is critical that the unit be mounted exactly as described in the proper installation techniques section when mounted horizontally. The exhaust outlet is located near the edge of the center (small) panel.

4. The entire heat exchanger assembly is mounted on rubber isolation mounts to eliminate vibration.

5. Each unit input size uses a specific heat exchanger assembly. Externally they are the same physical size and shape *but they must not be interchanged* between unit input sizes. Internal characteristics related to unit input properly match each assembly for the unit input rating. Refer to the

heat exchanger replacement kit installation instructions for replacement procedures.

> **Note:** If the heat exchanger must be replaced, be sure to keep the orifice from the old heat exchanger. New orifices are not shipped with new heat exchangers.

Spark Plug and Sensor

The following is a description of the spark plug and sensor:

1. The spark plug and sensor are located on the lower left side of the combustion chamber opposite the gas intake (see Figure 4.15). The sensor is the top plug (nearest to the blower) and is longer than the spark plug. The spark plug is in the lower position (farthest from the blower). The plugs cannot be interchanged due to the different thread diameters.
2. The spark plug socket size is three-quarters in. The sensor socket size is eleven-sixteenths in.
3. The spark plug is used in conjunction with the primary control for igniting the initial gas/air mixture.

> **Note:** On units using the Lennox GS-1 ignition control, a special resistor type plug *must be used* to prevent electrical interference from feeding back into the ignition control.

   The temperature in the combustion chamber keeps the plug free from oxides and it should not need regular maintenance. Compression rings are used to form the seal to the chamber.

4. Figure 4.16 gives the proper spark gap setting. Note that the spark plug used in the GSR14 uses a ground strap angle unusual in comparison to other spark plug applications. A feeler gauge can be used to check the gap.
5. The sensor is a spark plug type with a single center electrode (no ground strap). Compression rings are used to form the seal to the chamber. It should not need regular maintenance.

Gas and Air Components Applied to Heat Exchanger

The following is a description of these components. Figure 4.17 identifies all of the components that make up the basic heating assembly.

1. Gas flows through the valve, expansion tank, flapper valve, and the orifice into the combustion chamber.

IT IS NORMAL FOR THE GROUND
STRAP TO PROTRUDE AT AN UN-
USUAL ANGLE

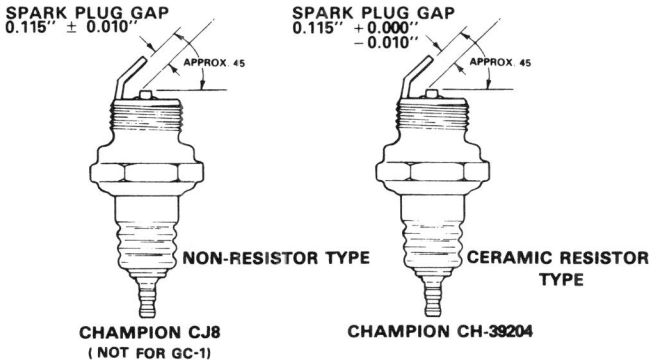

SPARK PLUG GAP
0.115" ± 0.010"

APPROX. 45

SPARK PLUG GAP
0.115" + 0.000"
       − 0.010"

APPROX. 45

NON-RESISTOR TYPE

CERAMIC RESISTOR
TYPE

CHAMPION CJ8
(NOT FOR GC-1)

CHAMPION CH-39204

NOTE-CARBON RESISTOR TYPE PLUGS SHOULD NOT BE USED

Figure 4.16 Spark plug gap setting.
(Courtesy of Lennox Industries, Inc.)

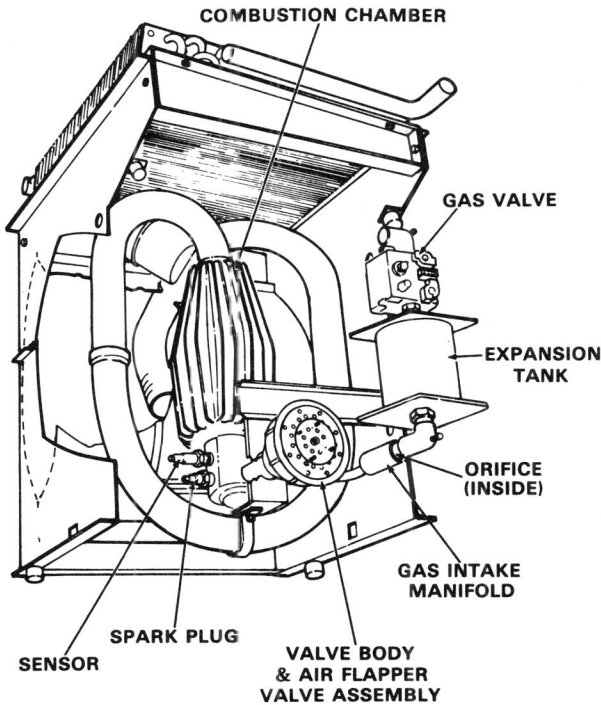

COMBUSTION CHAMBER

GAS VALVE

EXPANSION
TANK

ORIFICE
(INSIDE)

GAS INTAKE
MANIFOLD

SPARK PLUG

SENSOR

VALVE BODY
& AIR FLAPPER
VALVE ASSEMBLY

Figure 4.17 Gas and air components
applied to heat exchanger. (Courtesy
of Lennox Industries, Inc.)

2. Air flows through the flapper valve and directly into the combustion chamber.

3. Combustion takes place and the exhaust gas flows through the tailpipe, exhaust decoupler, and condenser coil to the exhaust outlet.

Blower Motors and Capacitors

All models use 120VAC PSC single-phase electric blower motors with RUN capacitors for efficiency (see Table 4.1).

TABLE 4.1 BLOWER MOTORS AND CAPACITORS (COURTESY OF LENNOX INDUSTRIES, INC.)

Models	HP	Capacitor MFD	Capacitor VAC
Q3	1/3	5	370
Q4	1/2	7	370
Q4/5	3/4	40	370

GSR14's use multitap blower motors (see blower speed, Table 4.2). Each motor is factory wired using the black high-speed tap for cooling and the red low-speed tap for heating (the GSR1403–80 uses the yellow medium-low-speed tap for heating).

TABLE 4.2 BLOWER SPEED SELECTION (COURTESY OF LENNOX INDUSTRIES, INC.)

Speed	Blower motor lead		
	Q3	Q4	Q4/5
Low	Red	Red	Red
Medium Low	Yellow	Yellow	Yellow
Medium			Blue
Medium Hi	Brown	Brown	Brown
High	Black	Black	Black

Important—to prevent motor burnout, never connect more than one motor lead to any one connection. Tape unused motor leads separately.

SEQUENCE OF OPERATION

For the proper sequence of operation refer to Figure 4.18.

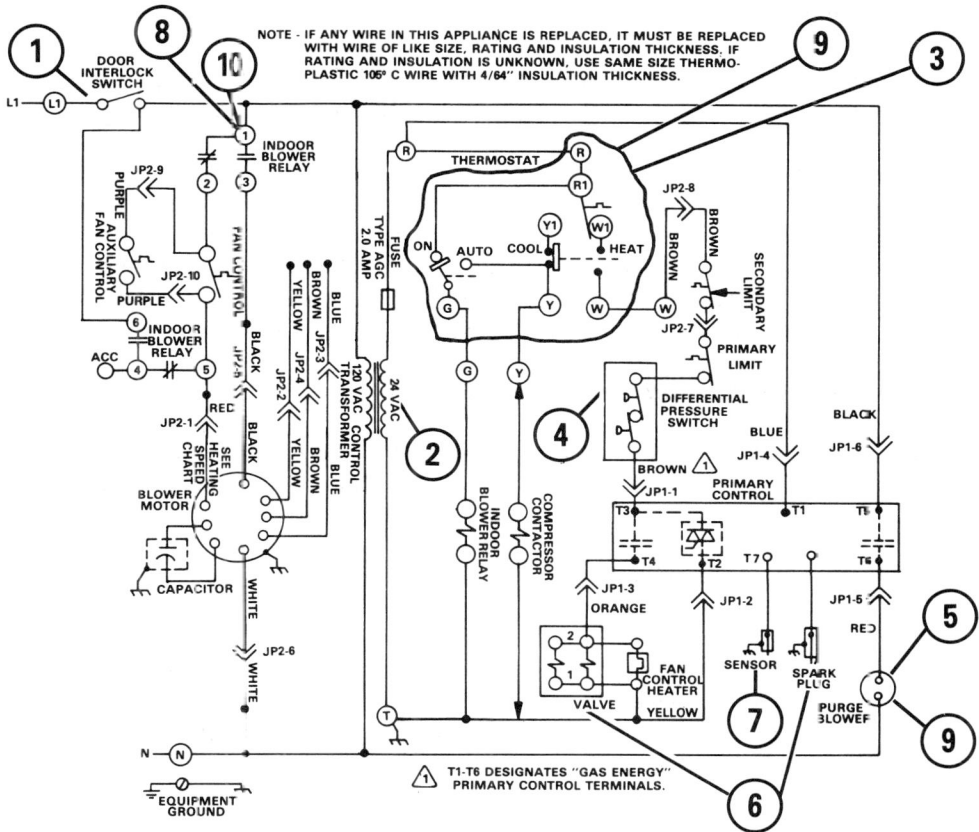

NOTE - IF ANY WIRE IN THIS APPLIANCE IS REPLACED, IT MUST BE REPLACED WITH WIRE OF LIKE SIZE, RATING AND INSULATION THICKNESS. IF RATING AND INSULATION IS UNKNOWN, USE SAME SIZE THERMO-PLASTIC 105° C WIRE WITH 4/64" INSULATION THICKNESS.

1 - Line voltage feeds through the door interlock switch. The blower access panel must be in place to energize unit.

2 - Transformer provides 24 volt control circuit power.

3 - A heating demand closes the thermostat heating bulb contacts.

4 - The control circuit feeds from 'W' leg through the secondary limit, the primary limit and the differential pressure switch to energize the primary control.

5 - Through the primary control the purge blower is energized for prepurge (see Figure 8 for length of prepurge).

6 - At the end of prepurge the purge blower continues to run and the gas valve, fan control heater and spark plug are energized (see Figure 8 for start attempt timings).

7 - The sensor determines ignition by flame rectification and de-energizes the spark plug and purge blower. Combustion continues.

8 - After approximately 30 to 45 seconds the fan control has warmed enough to close the contacts. The closed contacts energize the indoor blower motor on heating speed. If, at any time during a heating cycle, combustion is lost and heating demand is still present, the ignition control immediately returns to step 5 above.

9 - When heating demand is satisfied the thermostat contacts open. The primary control is then de-energized removing power from the gas valve and fan control heater. At this time the purge blower is energized for postpurge. The indoor blower motor remains on.

10 - When the air temperature lowers to 90°F the fan control contacts open - shutting the indoor blower off.

Figure 4.18 Sequence of operation. (Courtesy of Lennox Industries, Inc.)

INSTALLATION

Please refer to Lennox GSR14 series units Installation–Operation–Maintenance Instruction Manual for complete installation instructions.

Transporting the Unit

When moving or lifting the unit all access panels must be in place to prevent damage (sagging) to the unit. The unit should always be transported in the downflow position with the wooden shipping base installed. The blower may be removed to reduce the unit's weight while moving.

Support Frame and Suspension Rods

The unit must not be suspended by itself. A support frame must be used to prevent damage (sagging) to the unit. A support kit frame is available from Lennox.

Horizontal Mounting

When mounting the unit horizontally *it must be placed so that the air flow is from right to left.* This placement is necessary so that moisture can drain from the condensing coil *and must not be changed.* If mounted in any other position (such as inverted or on its back), the condensing coil will fill with condensate and make the unit inoperable (see Figure 4.19).

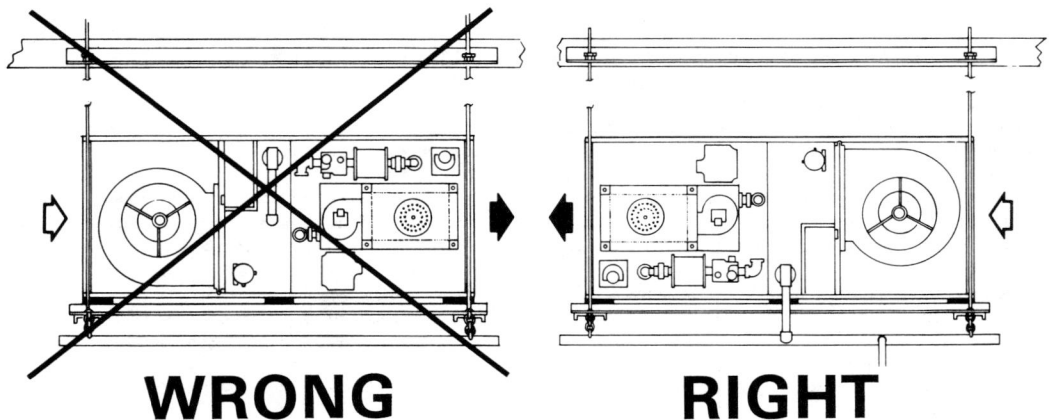

WRONG   RIGHT

Figure 4.19 Mounting the unit in a horizontal position. (Courtesy of Lennox Industries, Inc.)

Muffler

All CGA units and the AGA GSR14-100 units are shipped with an intake muffler and an exhaust line muffler. It is required that both be used. The mufflers are optional on AGA GSR14-50, 80 series units. *Any time an exhaust muffler is used in areas subjected to freezing temperature and especially if the muffler is mounted horizontally, a heat cable must be used on the muffler to prevent condensate from freezing inside.* Use only Lennox heat cable kits which are approved for use on PVC pipe.

Drip Leg and Condensate Line

When installing the unit in areas subjected to freezing temperature, the drip leg and condensate line *must* be wrapped with an electrical heat cable to prevent the condensate from freezing. A heat cable kit that is approved for use with PVC pipe is available from Lennox. Refer to GSR14 Heat Cable Kit Installation Instructions for proper application (see Figures 4.20 and 4.21).

Some installations may require a remotely mounted drip leg. Refer to GSR14 installation instructions for applications requiring remotely mounted drip leg and condensate line. The condensate line must be sloped away from the unit to allow proper drainage; a sagging line can hinder drainage.

Isolation Mounting Pads

Vibration mounting pads should be used, especially when the unit is installed on wood flooring. Isomode pads or equivalent should be used.

Flexible Boot Supply Air Plenum

A flexible canvas boot or equivalent is recommended but not required in the supply air plenum, downstream of the cooling coil, or future coil location.

Flexible Boot Return Air Plenum

A flexible boot or equivalent is recommended but not required in the return air plenum. It should be located as close to the furnace as possible, preferably between the furnace and external electronic air cleaner if used.

Gas Connector

(Used on AGA units only.) See Figure 4.22.

ALUMINUM FOIL TAPE

1″ INSULATION

HEAT CABLE (1 Wrap per Foot)

FIBERGLASS TAPE (Wrap around pipe 1-1/2 times)

PVC OR CPVC PIPE

HEAT CABLE WRAP PROCEDURE

HEAT CABLE

FOR 115V POWER SOURCE

WATER TIGHT
CONNECTOR

CLAMP

HANDY BOX
(Field Provided)

3 TO 4
INCHES

MOUNTING BRACKET

DRIP LEG

TYPICAL HEAT CABLE KIT INSTALLATION

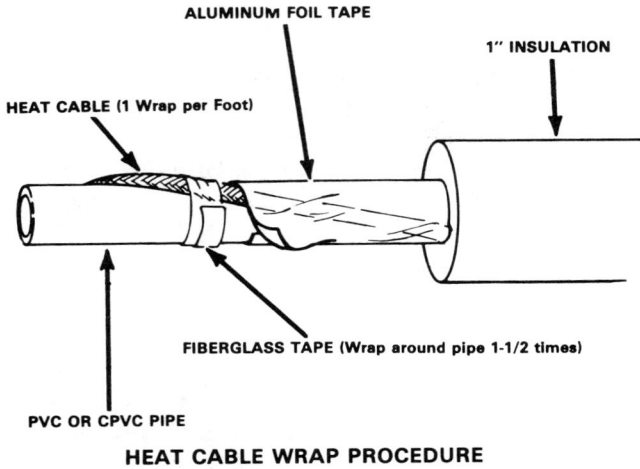

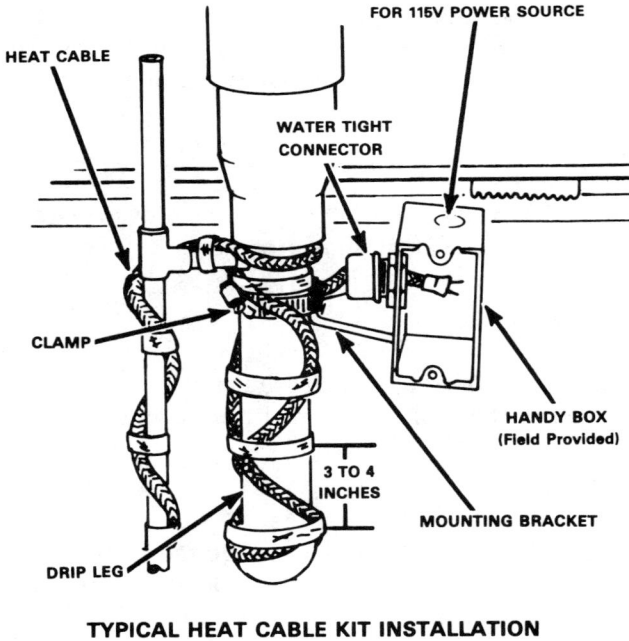

Figure 4.20 Typical heat wrap tape installation. (Courtesy of Lennox Industries, Inc.)

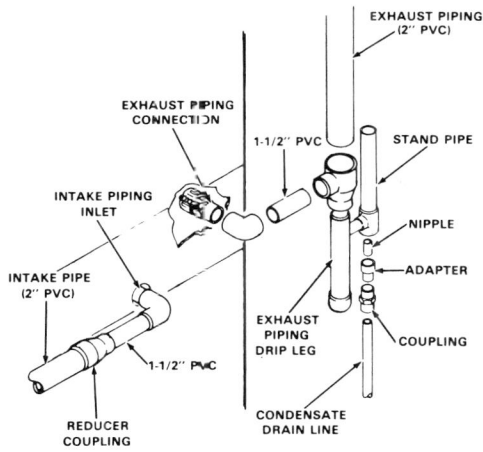

DOWNFLOW APPLICATION

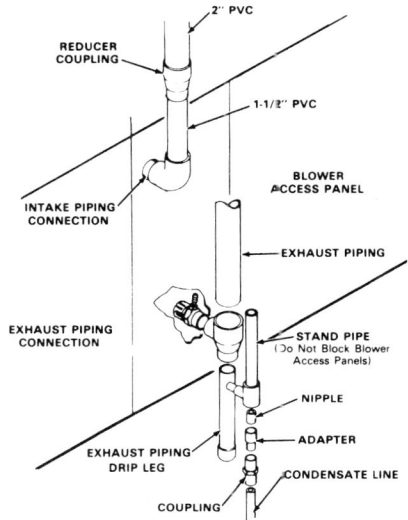

HORIZONTAL APPLICATION

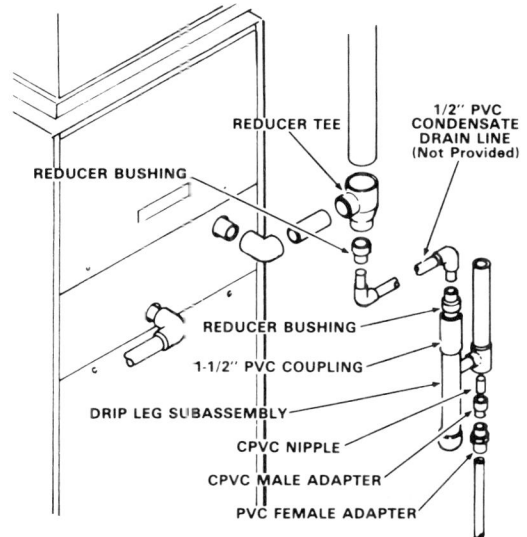

REMOTE INSTALLATION OF DRIP LEG

Figure 4.21 Typical drip leg installations. (Courtesy of Lennox Industries, Inc.)

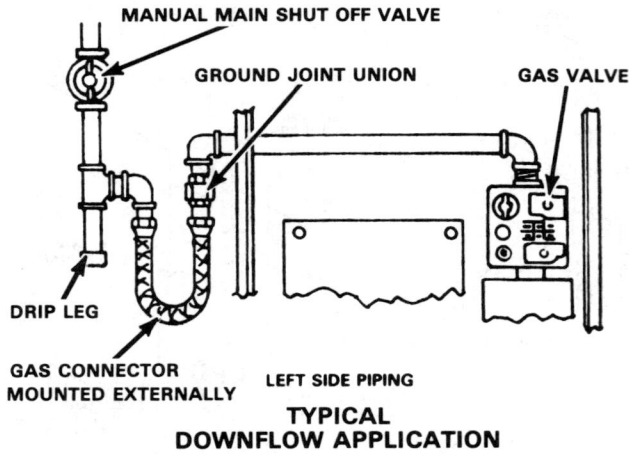

MANUAL MAIN SHUT OFF VALVE

GROUND JOINT UNION

GAS VALVE

DRIP LEG

GAS CONNECTOR MOUNTED EXTERNALLY

LEFT SIDE PIPING

TYPICAL DOWNFLOW APPLICATION

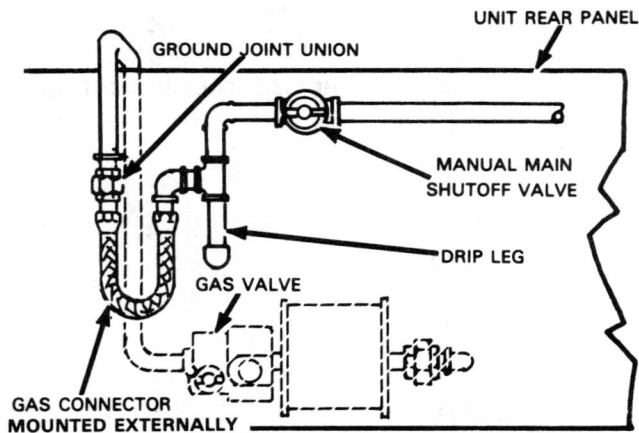

UNIT REAR PANEL

GROUND JOINT UNION

MANUAL MAIN SHUTOFF VALVE

DRIP LEG

GAS VALVE

GAS CONNECTOR MOUNTED EXTERNALLY

TYPICAL HORIZONTAL APPLICATION

Figure 4.22 Typical gas connection. (Courtesy of Lennox Industries, Inc.)

> **Caution:** The flexible gas connector (if used) *must* be mounted external to the unit. The connector must hang freely and must not rub against outside objects.

Gas Supply Piping Centered in Inlet Hole

The gas supply pipe should not rest on the unit cabinet.

Isolation Hangers

The PVC piping for the intake and exhaust line should be suspended from hangers. A suitable hanger can be fabricated from a 1-in.-wide strap of 26-gauge metal covered with Armaflex or equivalent.

Electrical Conduit Isolated from Ductwork and Joists

The electrical conduit can transmit vibration from the unit cabinet to ductwork or joists if clamped to either one. It may be clamped tightly to the cabinet but should not touch ductwork or joists.

Supply Air Plenum Insulated Past First Elbow

A 1.5- to 3-lb-density, matte face, 1-in.-thick insulation is required and all exposed edges should be protected from air flow.

Return Air Plenum Insulated Past First Elbow

A 1.5- to 3-lb-density, matte face, 1-in.-thick insulation is recommended but not required. All exposed edges should be protected from air flow.

Raised Platform

When mounting in a crawlspace or on attic beams it is important that the unit be supported by a flat base to prevent damage (sagging) to the unit.

Drain Pan

In any installation when the unit is mounted horizontally, a drain pan can be used to catch potential condensate leakage. The pan should be used on *all* applications where surrounding structures, such as walls or ceilings, might be damaged by potential condensate leakage. If an evaporator coil is used, the drain pan should be extended under the coil to catch potential condensate leakage.

Field Wiring

Field wiring is connected to terminal strips. Multispeed blower motors are factory wired with low-speed (red) taps for heating and high-speed (black) taps for cooling (GSR1403-80 uses the yellow medium-low-speed tap for heating) (see Figure 4.23).

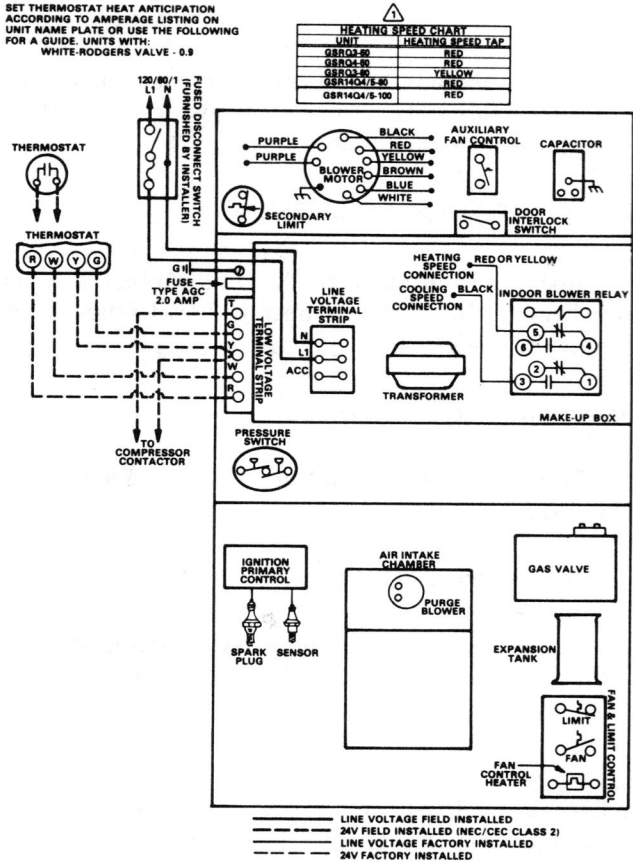

Figure 4.23 Field wiring diagram.
(Courtesy of Lennox Industries, Inc.)

Intake Exhaust Terminations

The intake and exhaust pipes should be placed as close together as possible at the termination end (see Figures 4.24–4.26.) *Maximum separation is 3 in. on roof terminations and 6 in. on side wall terminations. The end of the exhaust pipe must extend at least 8 in. past the end of the intake pipe.* The intake termination *must* be upwind (prevailing wind) of the exhaust pipe. Both intake and exhaust *must* be in the same pressure zone. (Do not exit one through the roof and one through the side of the house.) These precautions are to ensure that exhaust gas recirculation does not occur.

The exhaust piping must terminate straight out or up and the termination must not be within 6 ft of other vents or 3 ft of structure openings.

Refer to the GSR14 Operation and Installation Instructions for detailed instructions of proper termination installations that meet local and national codes.

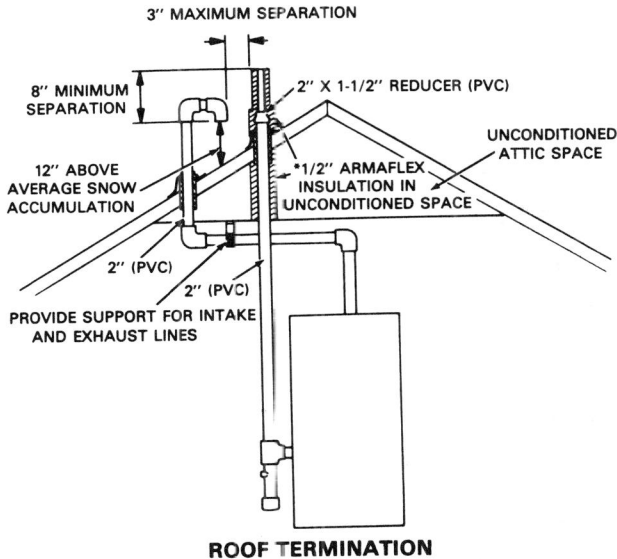

ROOF TERMINATION

*IF WINTER DESIGN TEMPERATURE IS BELOW 32°F, 1/2" ARMAFLEX INSULATION SHOULD BE USED IN EXTREME COLD CLIMATE AREAS. 3/4" ARMAFLEX INSULATION RECOMMENDED.

Figure 4.24 AGA and CGA units roof termination. (Courtesy of Lennox Industries, Inc.)

Exhaust Insulation

In areas subject to freezing temperature, the exhaust pipe must be insulated with one-half in. Armaflex or equivalent when run through an unheated space.

MAINTENANCE

At the beginning of each heating season, the system should be checked as follows:

Supply Air Blower

Use the following steps to check the supply air blower:

1. Check and clean the blower wheel.
2. Always lubricate the blower motor according to the manufacturer's lubrication instructions on each motor. If no instructions are provided, use the following as a guide.
   a. Motors without oiling ports, prelubricated and sealed: No further lubrication is required.

12" MAXIMUM

NOTE - REDUCE TO 1-1/2"

1-1/2" (PVC)   EXHAUST

UNCONDITIONED SPACE

INTAKE   8" MINIMUM

OUTSIDE WALL

*1/2" ARMAFLEX
INSULATION

PROVIDE SUPPORT FOR
INTAKE AND EXHAUST
LINES EVERY 3 FEET

12" ABOVE AVERAGE
SNOW ACCUMULATION

*1/2" ARMAFLEX
INSULATION IN
UNCONDITIONED SPACE

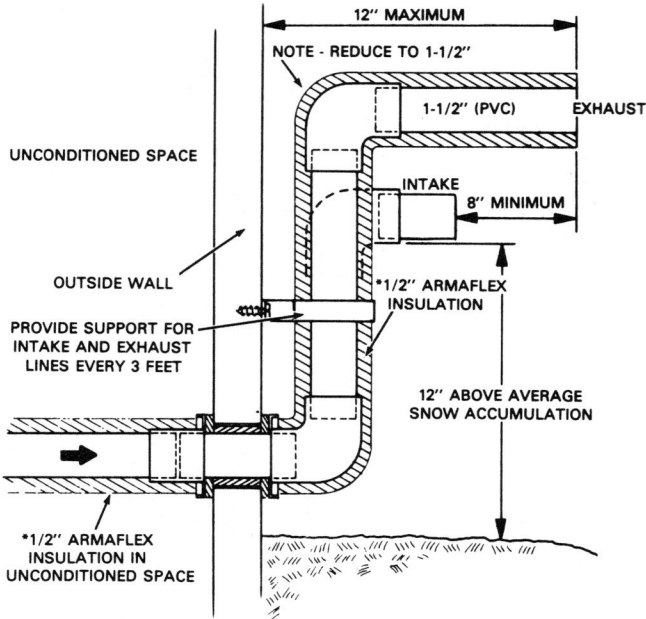

EXHAUST LINE WALL TERMINATION

*IF WINTER DESIGN TEMPERATURE IS BELOW 32°F, 1/2" ARMAFLEX INSULATION
SHOULD BE USED IN EXTREME COLD CLIMATE AREAS. 3/4" ARMAFLEX
INSULATION RECOMMENDED. 12" MAXIMUM

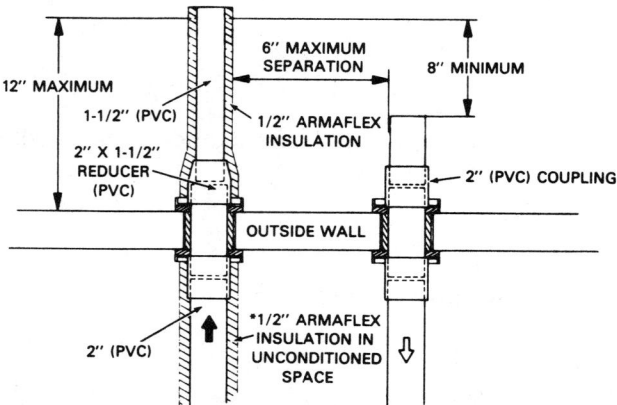

6" MAXIMUM
SEPARATION   8" MINIMUM

12" MAXIMUM

1-1/2" (PVC)   1/2" ARMAFLEX
INSULATION

2" X 1-1/2"
REDUCER   2" (PVC) COUPLING
(PVC)

OUTSIDE WALL

*1/2" ARMAFLEX
2" (PVC)   INSULATION IN
UNCONDITIONED
SPACE

TOP VIEW WALL TERMINATION

*IF WINTER DESIGN TEMPERATURE IS BELOW 32°F, 1/2" ARMAFLEX INSULATION
SHOULD BE USED IN EXTREME COLD CLIMATE AREAS. 3/4" ARMAFLEX
INSULATION RECOMMENDED.

Figure 4.25 CGA units only, individual wall terminations. (Courtesy of Lennox Industries, Inc.)

NOTE - Kit is designed to penetrate 12 inches inside wall. If necessary use couplings to add extra length.

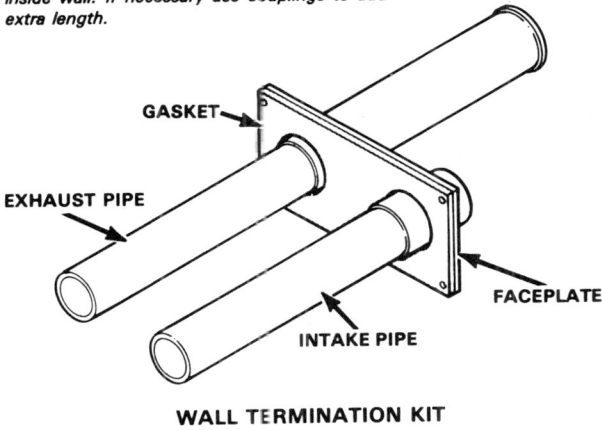

GASKET

EXHAUST PIPE

FACEPLATE

INTAKE PIPE

WALL TERMINATION KIT

1/2" Armaflex insulation in unconditioned space

OUTSIDE WALL

TOP VIEW
WALL TERMINATION

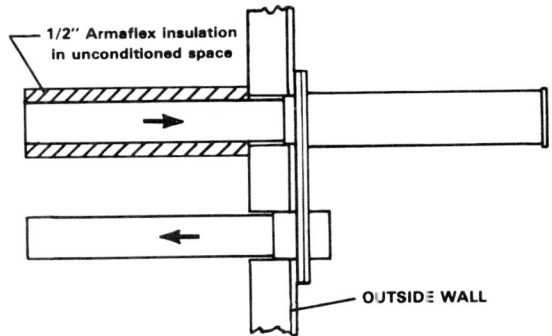

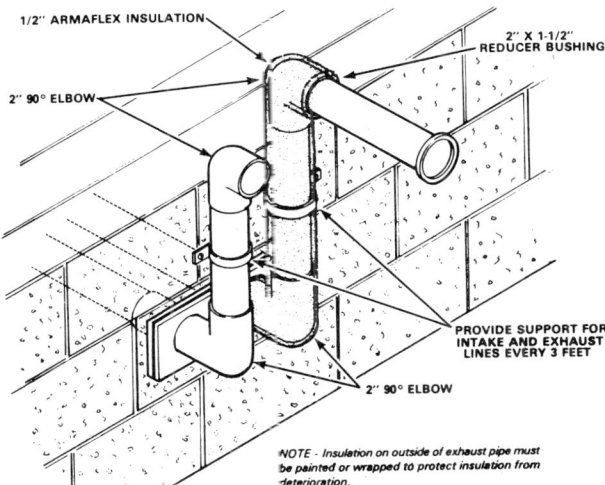

1/2" ARMAFLEX INSULATION

2" X 1-1/2" REDUCER BUSHING

2" 90° ELBOW

PROVIDE SUPPORT FOR INTAKE AND EXHAUST LINES EVERY 3 FEET

2" 90° ELBOW

NOTE - Insulation on outside of exhaust pipe must be painted or wrapped to protect insulation from deterioration.

NOTE - Insulation on outside runs of exhaust pipe must be painted or wrapped to protect insulation from deterioration.

Figure 4.26 AGA units only, wall termination kit. (Courtesy of Lennox Industries, Inc.)

b. Direct drive motors with oiling ports, prelubricated for an extended period of operation: For extended bearing life relubricate with a few drops of SAW 10W nondetergent oil once every two years. It may be necessary to remove the blower assembly for access to the oiling ports.

Filters

Use the following steps to check the filters:

1. The filters must be cleaned or replaced when dirty to assure proper unit operation. Replace with 20 × 25 × 1-in. filters.
2. The filters supplied with the GSR14 can be washed with water and a mild detergent. They should be sprayed with Filter Handicoater when dry prior to reinstalling in the unit. The Filter Handicoater is RP Products coating No. 418 and is available as Lennox part No. P–8-5069.

Fan and Limit Controls

Check fan and limit controls for proper operation and (where possible) setting. For settings, refer to the Fan and Limit section of this chapter.

Electrical

Use the following steps to check the electrical parts of the furnace:

1. Check all wiring for loose connections.
2. Check for the correct voltage.
3. Check the amperage draw on the blower motor.

Intake and Exhaust Lines

Use the following step to check the intake and exhaust lines:

1. Check the intake and exhaust PVC lines and all connections for tightness and make sure that there is no blockage. Also, check the condensate line for free-flowing operation and complete drainage.

TYPICAL OPERATING INSTRUCTIONS

The following are the typical operating characteristics of the Lennox GSR14 Unit:

Temperature Rise

The temperature rise of the GSR14 unit depends on the unit input, blower speed, blower horsepower, and static pressure as marked on the rating plate of the furnace. The blower speed must be set for unit operation within the range of AIR TEMPERATURE RISE °F listed on the unit rating plate.

To measure temperature rise. Use the following steps to measure the temperature rise of the unit:

1. Place plenum thermometers in the warm air and the return air plenums. Locate the thermometer in the warm air plenum where it will not "see" the heat exchanger, thus picking up radiant heat.
2. Set the room thermostat to the highest setting.
3. After the plenum thermometers have reached their highest reading, subtract the two readings. The difference should be in the range listed on the unit rating plate. If this temperature is low, decrease the blower speed; if it is high, increase the blower speed. To change the blower speed, refer to Table 4.2.
4. Be sure to check and determine that the discharge static pressure is within the range of values listed on the unit rating plate before adjusting the blower speed.

To measure static pressure. Use the following steps to measure the static pressure of an air distribution system:

1. Measure the tap locations as shown in Figure 4.27.
2. Punch a one-fourth-in.-diameter hole in the plenum. Insert the manometer hose flush with the inside edge of the hole or insulation. Seal around the hose with permagum. Connect the zero end of the manometer to the discharge (supply) side of the plenum. On ducted systems, connect the other end of the manometer to the return duct as above. For systems with nonducted returns, leave the other end of the manometer open to the atmosphere.
3. With only the blower running, observe the manometer reading. Adjust the motor speed to deliver the air desired according to the unit rating plate.
4. Seal the hole when the check is completed.

Manifold Pressure

Checks of the manifold pressure are made as verification of proper regulator adjustment. The manifold pressure for the GSR14 can be made any time the

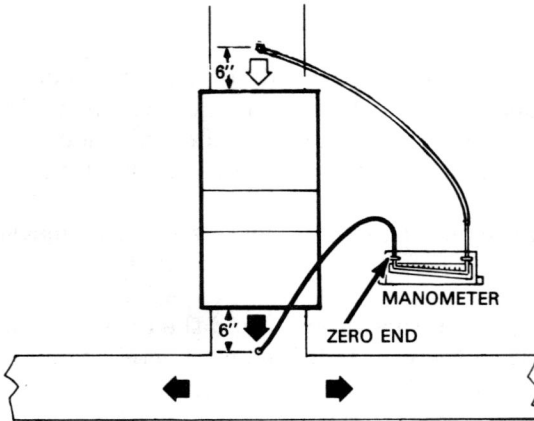

1/4" DIAMETER HOLE CENTERED ACROSS SUPPLY PLENUM
OPEN TO ATMOSPHERE ON NON-DUCTED SYSTEMS, FOR DUCTED
SYSTEMS CONNECT RETURN AIR DUCT SIDE

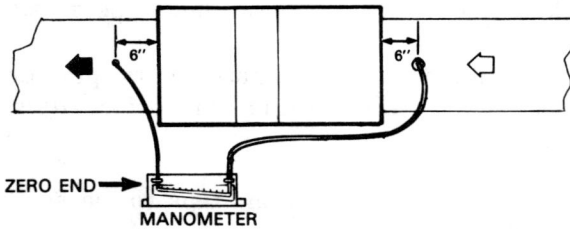

Figure 4.27 Measuring discharge static pressure on a furnace. (Courtesy of Lennox Industries, Inc.)

gas valve is open and is supplying gas to the unit. Normal manifold pressure is 2.0 ± 0.2 in. wc for natural gas and 9.0 ± 0.2 in. wc for LP gas.

To measure.   Use the following steps to check the manifold pressure:

1. Remove the one-eighth-in. pipe plug from the pressure tap on the elbow below the expansion tank (see Figure 4.2).

> **Caution:** For safety, connect a shutoff valve between the manometer and the gas tap to permit shutoff of the gas pressure to the manometer, if desired.

2. Insert a hose adapter in the tap and connect the gauge.
3. Set the room thermostat for heating demand (the demand can be started at the unit by jumping R to W on the low-voltage terminal strip). *Be sure to remove the jumper after the test is complete.*

4. Check the manifold pressure after the unit has ignited and is operating normally. See step 5 for units that will not ignite.

5. If the unit is not operational, *see cautions and warnings below.* Check the manifold pressure immediately after the gas valve opens fully.

Caution: Disconnect the heating demand as soon as an accurate reading is obtained and allow the unit to postpurge the heat exchanger before proceeding.

Warning *The combustion chamber access panel and the air decoupler box cover must be in place for this test. Do not allow long periods of trial for ignition. Unsafe conditions could result.*

Warning: *If the unit is not operational, the manifold pressure check should be used only to verify that gas is flowing to the combustion chamber at the correct manifold pressure. Always allow the purge blower to evacuate the combustion chamber before proceeding.*

---

6. The gas valve can be adjusted using the regulator adjustment screw. This screw is located under the dust cover screw on the face of the valve next to the electrical terminals and the manual on/off knob (refer to Figure 4.12).

7. Always check the manifold pressure after adjusting the regulator. Turn the adjustment screw clockwise to increase the manifold pressure and counterclockwise to reduce the manifold pressure.

Line Pressure

The gas supply pressure should not exceed 13.0 in. wc and should not drop below 3.5 in. wc. The supply pressure should be checked only with the unit running. A one-eighth-in. pipe plug and tap are supplied in the elbow on the inlet side of the gas valve for checking line pressure. Line pressure ratings are also listed on the unit rating plate.

Flame Signal

A 50-μA dc meter is needed to check the flame signal on the ignition primary controls. (Simpson models 250, 255, and 260 have a dc range of 0–50 μA and are suggested for use.)

To measure.   Use the following steps to measure the line gas pressure:

1. Place the meter in series between the ignition control and the sensor wire; positive (+) lead of the meter to the ignition control sensor connection and the negative (−) lead of the meter to the sensor wire.
2. Set the room thermostat for a heating demand and check the flame signal with the unit operating.
3. The flame signals are:
   Lennox GC1 primary control: 1–25 μA dc
   Watsco primary control: 2–5 μA dc
   Gas Energy primary control: 18–35 μA dc
   The flame signal may rise above these values for the first few seconds after ignition and then level off within the ranges listed above.

Exhaust Temperature Range

The exhaust temperature should not exceed 135 °F for any of the GSR14's. If the exhaust temperature exceeds 135 °F, the auxiliary fan control may activate and either the primary limit or the secondary limit may "trip." A high exhaust temperature may indicate a problem in the unit or restrictions in the ducts or PVC pipes.

Most units run with a maximum exhaust temperature of 110 to 125 °F from the lower to the higher unit inputs. Exhaust temperatures lower than these are normal.

Exhaust CO_2 (Carbon Dioxide) Content

Carbon dioxide is a colorless odorless gas produced in small amounts by all furnaces, including the GSR14, during the combustion process. When the unit is properly installed and operating normally, the CO_2 content of the exhaust gas is within 8.0–10.0% for natural gas and within 9–11% for LP gases. If the unit appears to be operating normally at or beyond the upper limit of the CO_2 range, the unit should be checked for abnormally high CO (carbon monoxide) output, which might indicate other problems in the system.

One method of measuring the CO_2 content is to use the Bacharach CO_2 test with a Fyrite CO_2 indicator. Other methods of testing CO_2 are available. Closely follow the instructions included with the test kit you choose. A method for connecting the CO_2 test kit to the GSR14 is outlined later in this chapter.

Exhaust CO (Carbon Monoxide) Content

If the unit appears to be operating normally with a CO_2 output at or near the upper limits listed under Exhaust CO_2 (Carbon Dioxide) Content, the unit should be checked for an abnormally high CO content. When the unit is properly installed and operating normally, the CO content of the exhaust gas is less than 0.04%, regardless of the type of gas used. Conditions that cause abnormally high CO include but are not limited to:

1. Partial blockage of the exhaust pipe and intake pipe.
2. Abnormally high exhaust back pressure and intake restriction due to pipe length or routing.
3. The type of gas used.
4. Atmospheric conditions present at a particular location.

> **Warning:** *A high CO output may be fatal. Do not allow the unit to operate at CO output levels above 0.04%. Before allowing the unit to operate, the source of improper combustion must be located and corrected.*

To connect a device for taking CO_2 or CO samples. Use the following steps to check the CO_2 or the CO content of the exhaust gas:

1. Disconnect the differential pressure switch hose from the one-eighth-in. differential pressure switch outlet adaptor and connect the sampling device to the hose (see Figure 4.28).
2. Set the room thermostat to the highest setting and allow the unit to run for 15 minutes before taking a sample.
3. Take the CO_2 or the CO sample.
4. When the CO_2 or the CO test is completed, turn off the unit, remove the test hose from the pressure switch hose and *securely* reconnect the differential pressure switch hose to the pressure switch adaptor.

> **Caution:** *Make sure that the differential pressure switch hose is securely reconnected to the pressure switch adaptor.* The exhaust vent pipe operates under positive pressure and must be completely sealed to prevent leakage of combustion products into the living space.

Safety Shutdown

Safety shutdown occurs when any of the following problems are encountered:

1. Loss of combustion during the heating cycle caused by:
   a. Obstruction in the air intake piping.
   b. Obstruction in the exhaust piping.

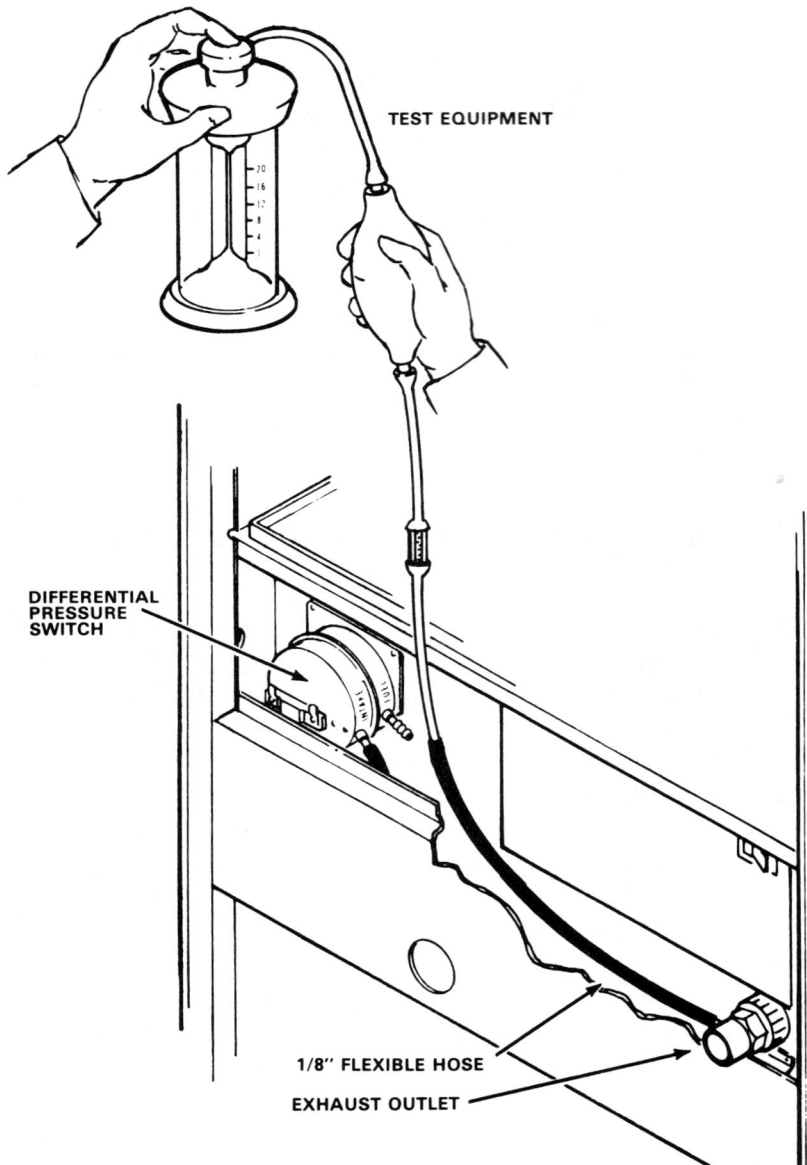

Figure 4.28 Connections for taking CO_2 and CO test samples. (Courtesy of Lennox Industries, Inc.)

    c. Low gas pressure.
    d. Failure of the gas flapper valve.
    e. Failure of the air flapper valve.
    f. Failure of the main gas valve.
    g. Loose spark plug or sensor creating a pressure loss in the combustion chamber.
    h. Loose sensor wire.
2. Primary limit cutout:
    a. Indoor blower failure.
    b. Temperature rise too high.
    c. Restricted filter or return air.
    d. Restricted discharge air.
3. Secondary limit cutout:
    a. Indoor blower failure.
    b. Temperature rise too high.
    c. Restricted filter or return air opening.
    d. Restricted discharge air.
    e. Primary fan control shutting the blower off too early.
    f. Auxiliary fan control by activating the blower is unable to compensate for the temperature rise in the blower housing.

Internal Component Temperatures

During operation, the temperature at the top of the combustion chamber and tail pipe is 1000–1200 °F. At the tailpipe entrance to the exhaust decoupler, the temperature has dropped to approximately 600 °F. From the exhaust decoupler outlet to the coil intake manifold the temperature has dropped to approximately 350 °F. At the coil exhaust outlet manifold the temperature ranges from approximately 100 to 110 °F. These are average temperatures and will vary with blower speed and gas input.

Condensate pH Range

The condensate is mildly acidic and can be measured with pH indicators. The pH scale is a measurement of acidity and alkalinity.

    Scale 4.1 shows the relative pH of some common liquids as compared with the condensate of GSR14 units. The concentration of the acidity of all these fluids including the condensate is very low and harmless.

```
↑                           0
                            1
                            2 - Vinegar
Increasing                  3 - Wine
Acidity
                            4 - Orange Juice          ⎫
                            5 - Tomato Juice          ⎬  GSR14 Condensate
                                                      ⎭  pH Range
                            6
─────────────────────────────────────────────────────────────
                            7 - tap water
─────────────────────────────────────────────────────────────
                            8
                            9
                            10
Increasing                  11
Alkalinity
                            12
                            13
↓                           14
─────────────────────────────────────────────────────────────
```

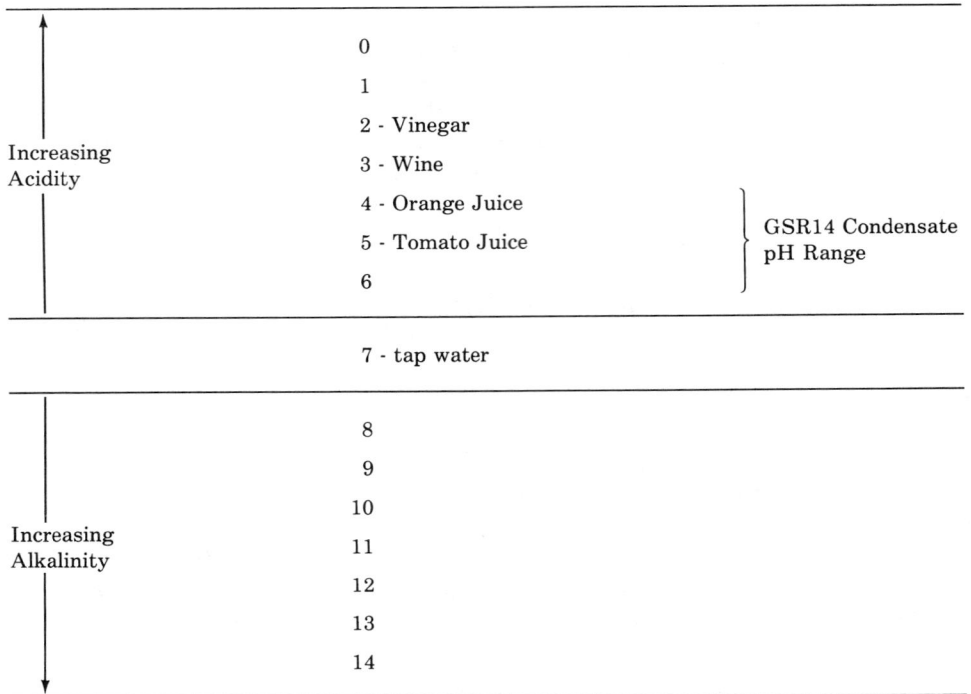

SCALE 4.1   RELATIVE pH OF SOME LIQUIDS. (COURTESY OF LENNOX INDUSTRIES, INC.)

Acceptable Operating Input

Field adjustments to the unit are not normally needed due to the specifically sized components for each unit input rating.

The unit may run up to ±3 to 4% of the rated input (listed on the unit nameplate) due to installation variables such as temperature rise, external static pressure, and return air temperature combined with allowable tolerances of components within the unit. This is an acceptable operating range.

Operation of the GSR14 above or below this acceptable operating range may cause continuity, start-up, and lockout problems (erratic operation). *Overfiring the unit can shorten the life of the heat exchange assembly.* Overfiring can be checked by measuring the input to the unit.

For new units, approximate input measurements may be obtained after allowing the unit to run continuously for 15 minutes. If accurate input measurements are required, the following procedure may be used. To achieve accurate input measurement requires time for the unit to "run in." The run-in time allows the flapper valves to "seat" and combustion to clean the protective layer of oil residue that may still be present from inside the heat exchange assembly. This process stabilizes the combustion rate and may take 1–2 hours of continuous operation. Since it is impractical to operate an installed unit for 1–2 hours continuously, the unit should be allowed to operate normally (cycle on demand) for a period to accumulate several total hours of run time. Overnight operation should provide enough total run time to obtain an accurate measurement of input. Just prior to any input check the unit should be run continuously for 15 minutes.

Checking gas input—determine gas flow at meter. Use the following steps for this procedure:

1. Turn off all other gas appliances, including pilot lights on appliances if used.
2. Set the room thermostat to the highest setting and allow the unit to run continuously for 15 minutes. The 15-minute run time is needed to allow the unit to stabilize the operating rate.
3. At the gas supply meter and using either the 1-, 2-, or 5-ft dial on the meter, time one full revolution (in seconds) with a stop watch (see Figure 4.29).
4. Find the number of seconds for one revolution on the Gas Rate Chart (see Table 4.3). Read the cubic feet for the matching 1-, 2-, or 5-ft dial size from the table and multiply this times the Btu per cubic foot content of the gas. The result is the total Btu input to the furnace.

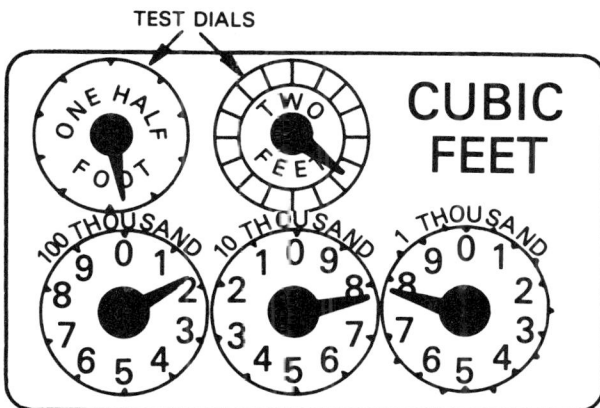

Figure 4.29 Gas meter dial. (Courtesy of Lennox Industries, Inc.)

TABLE 4.3 GAS RATE CHART (COURTESY OF LENNOX INDUSTRIES, INC.)

Meter flow rate

Gas rate—cubic feet per hour

	Size of test meter dial			
Secs. for one rev.	1/2 Cubic feet	1 Cubic feet	2 Cubic feet	5 Cubic feet
10	180	360	720	1,800
11	164	327	655	1,636
12	150	300	600	1,500
13	138	277	555	1,385
14	129	257	514	1,286
15	120	240	480	1,200
16	112	225	450	1,125
17	106	212	424	1,059
18	100	200	400	1,000
19	95	189	379	947
20	90	180	360	900
21	86	171	343	857
22	82	164	327	818
23	78	157	313	783
24	75	150	300	750
25	72	144	288	720
26	69	138	277	692
27	67	133	267	667
28	64	129	257	643
29	62	124	248	621
30	60	120	240	600
31	58	116	232	581
32	56	113	225	563
33	55	109	218	545
34	53	106	212	529
35	51	103	206	514
36	50	100	200	500
37	49	97	195	486
38	47	95	189	474
39	46	92	185	462
40	45	90	180	450
41	44	88	176	440
42	43	86	172	430
43	42	84	167	420
44	41	82	164	410
45	40	80	160	400
46	39	78	157	391
47	38	77	153	383
48	37	75	150	375
49	37	73	147	367
50	36	72	144	360
51	35	71	141	353

TABLE 4.3 GAS RATE CHART (COURTESY OF LENNOX INDUSTRIES, INC.)

Meter flow rate

Gas rate—cubic feet per hour

Secs. for one rev.	Size of test meter dial			
	1/2 Cubic feet	1 Cubic feet	2 Cubic feet	5 Cubic feet
52	35	69	138	346
53	34	68	136	340
54	33	67	133	333
55	33	65	131	327
56	32	64	129	321
57	32	63	126	316
58	31	62	124	310
59	30	61	122	305
60	30	60	120	300
62	29	58	116	290
64	29	56	112	281
66	29	54	109	273
68	28	53	106	265
70	26	51	103	257
72	25	50	100	250
74	24	48	97	243
76	24	47	95	237
78	23	46	92	231
80	22	45	90	225
82	22	44	88	220
84	21	43	86	214
86	21	42	84	209
88	20	41	82	205
90	20	40	80	200
94	19	38	76	192
98	18	37	74	184
100	18	36	72	180
104	17	35	69	173
108	17	33	67	167
112	16	32	64	161
116	15	31	62	155
120	15	30	60	150
130	14	28	55	138
140	13	26	51	129
150	12	24	48	120
160	11	22	45	112
170	11	21	42	106
180	10	20	40	100
190	9.5	19	38	95
200	9	18	36	90
210	8.5	17	34	86
220	8	16	33	82

Example:

a. One revolution on the 2-ft dial = 90 seconds.

b. Using the Gas Rate Chart, Table 4.3, note that 90 seconds = 80 ft³ of gas per hour.

c. Nominally there are 1000 Btu in each cubic foot of natural gas. Make adjustment to this figure where the gas heating value is other than 1000 Btu per cubic foot (contact the local gas supplier for local Btu per cubic foot gas ratings).

d. 80 ft³ per hour × Btu per cubic foot content = 80 000 Btuh input.

5. Check the Btuh input figure against the Btuh listed on the unit nameplate.

NEW UNIT START-UP

Normal start-up conditions of a new installation may require running the unit through several tries for ignition before the unit will run continuously. Initially, the unit may start and die several times until all the air bleeds from the gas piping and residues of oil and water are purged from the heat exchange assembly. Break and remake the room thermostat demand to restart the ignition sequence at 2- to 3-minute intervals until continuous operation is obtained.

TROUBLESHOOTING

Troubleshooting the GSR14 depends on complete understanding of how all the components work, as described in this chapter. Common problems are broken down into four main categories:

1. Unit will not run.
2. Unit starts clean but runs less than 10 seconds.
3. Unit starts but shuts off before the room thermostat is satisfied— insufficient heat.
4. Unit sputter starts and dies.

Each of the four problem categories above are broken down into troubleshooting flow charts in the last pages of this chapter with additional information provided to explain certain checks. Steps in the flowcharts for measuring manifold pressure, flame signal, exhaust CO_2 content, exhaust CO content, and operating input are explained in previous sections.

Choose the flowchart that best describes the unit's problem. Follow the flowchart step by step. At any point that a NO answer is reached and a repair is made, reassemble the unit and retest for operation. If the unit does not operate, recheck up to that point and then continue through the chart. Occasionally more than one problem may exist.

When troubleshooting a unit be sure that all of the basic checks are covered carefully and double check your diagnosis before replacing components. *Do as little disassembly as possible* during troubleshooting to prevent introducing additional problems such as gas or air leaks or damage to components.

Caution: Before servicing the unit:

1. If the unit has been operating, the internal components will be *hot.* Allow the unit to cool for at least 15 minutes before placing your hands into the heat section access opening.

2. To cool the unit completely to room temperature, the blower should run continuously for about 40 minutes.

3. When servicing the air intake flapper valve, keep in mind that it is only moderately warm during unit operation. After the unit cycles off, the residual heat in the combustion chamber will transfer back to the valve causing it to become *very hot!* Allow it to cool for 10 to 15 minutes before handling. The blower can also be run to cool the air intake.

4. The spark plug is torqued to 28 ± 2 ft-lb. The sensor is torqued to 14 ± 1 ft-lb.

Allow the metal to cool before measuring torque.

Checking Air Intake Flapper Valve

Use the following procedure for checking the air intake flapper valve:

a. Remove the air intake chamber cover and check for foreign materials that may have accumulated. Clean and purge the blower and upper and lower chamber compartments if necessary.

b. Do not remove the air flapper valve unless it is suspected of being faulty. If the air flapper valve is removed, new screws must be used to resecure the valve to the valve body.

> **Note:** The special screws used with the air flapper valve are treated with a nonhardening seizing compound. *Do not reuse old screws. New screws are supplied with new replacement air flapper valves.* If replacing the old air flapper valve in the unit, *new replacement screws are available in kit form from Lennox Repair Parts.* Do not use Loctite or similar thread seizing compounds to secure old screws.

If the valve must be removed, carefully remove the eight screws holding the air intake flapper valve to the body. *Do not turn or remove the center screw.* Remove the flapper valve from the unit, being careful not to damage the gasket (if used, refer to items e and f below). **Caution:** *Do not drop.*

c. *Do not disassemble the internal components of the valve.* If taken apart, the plates can be rotated out of phase or reversed and the spacer thickness has an extremely low tolerance. Note that each plate has a stamp of the spacer thickness, a star, or the words THIS SIDE OUT stamped on it. These stamps should all lie in the same quadrant and face the outside of the unit (see Figure 4.30).

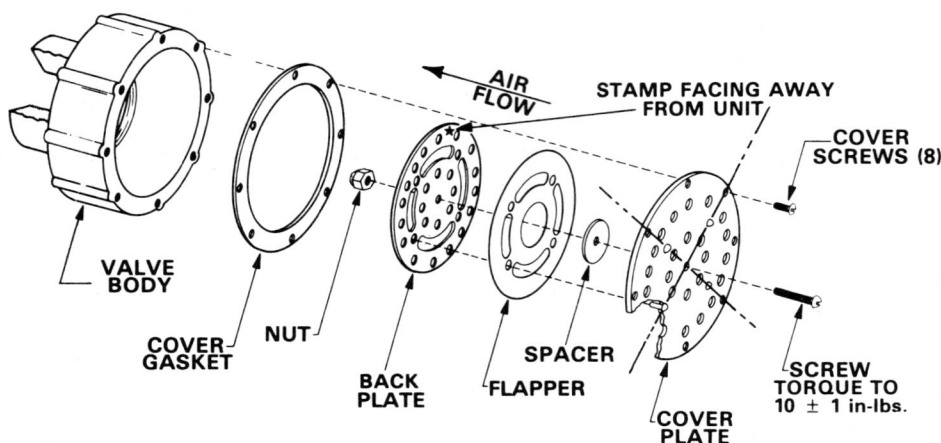

Figure 4.30 Air flapper valve assembly (50's and 80's only). (Courtesy of Lennox Industries, Inc.)

d. Visually inspect the flapper. On new units, the flapper may not be perfectly flat: It may be curved or dished between the plates. This is normal. On units that have had sufficient run-in time, the flapper will be flat. If the flapper is torn, creased, or has uneven (frayed) edges, the valve assembly must be replaced.

e. For models Q3, 4-50, 80 only: *The air flapper valves for these units require a gasket between the valve and the valve body* for proper operation (see Figure 4.30). To find potential warpage in the plates, check for the required clearance between the flapper and the back plate in several places around the circumference of the valve. Use a feeler gauge, starting small and working up to the clearance dimension until the gauge is just about snug. *Be very careful not to damage the flapper material by forcing the feeler gauge.* The clearance should be checked in 6 or 8 places around the valve. If the valve is out of the clearance dimension given in Figure 4.31 at any point around the valve, simply replace the assembly.

MODEL	FUEL	DIMENSION 'A' inches
GSR14-50 (all models)	Natural	0.025 ± 0.003
GSR14-50 (all models)	L.P.	0.030 ± 0.003
GSR14-80 (excluding CGA Hi-Alt.)	Natural	0.035 ± 0.003
GSR14-80 (excluding CGA Hi-Alt.)	L.P.	0.035 ± 0.003
GSR14-80 (CGA Hi-Alt.)	Natural	0.050 ± 0.003
GSR14-80 (CGA Hi-Alt.)	L.P	0.050 ± 0.003

NOTE - Some L.P. gas GSR14-50's, may be equipped with 0.025" ± 0.003" spacers.

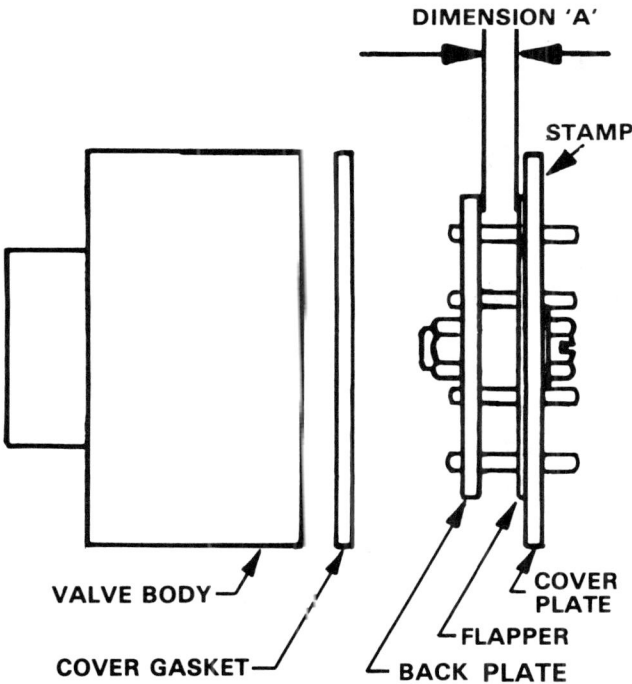

Figure 4.31 Air flapper valve required clearance. (Courtesy of Lennox Industries, Inc.)

f. For models Q4/5-100 only: *The flapper valve assembly for this unit does not require a gasket between the valve and the valve body* (see Figure 4.32). The air flapper assembly is designed as a one-piece unit. Clearances between the flapper and the back plate cannot be readily checked. If it is suspected of being faulty after a visual inspection, simply replace the assembly.

VALVE BODY

NO GASKET

STAMPS ALL IN SAME
QUADRANT AND FACING
AWAY FROM UNIT

CENTER
PLATE

BACK PLATE

DIMENSION 'A'

FLAPPER

QUADRANTS

TORQUE TO 10 ± 1 in-lbs.

COVER SCREWS (8)

DIMENSION 'A'

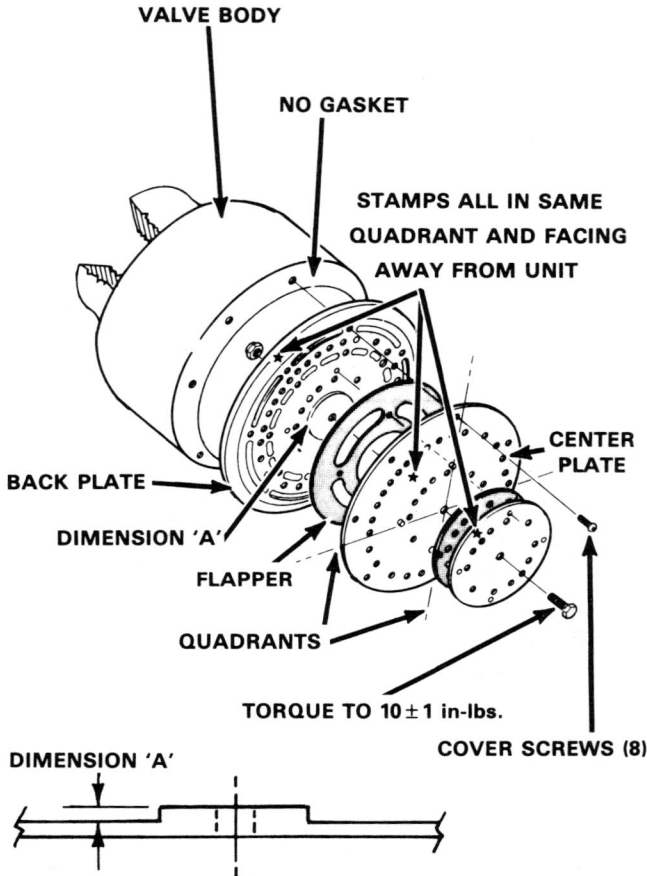

MODEL	FUEL	DIMENSION 'A' inches
GSR14-100 (excluding CGA Hi-Alt.)	Natural	0.040 ± 0.0015
GSR14-100 (excluding CGA Hi-Alt.)	L.P.	0.045 ± 0.0015
GSR14-100 (CGA Hi-Alt.)	Natural	0.060 ± 0.0015
GSR14-100 (CGA Hi-Alt.)	L.P.	0.060 ± 0.0015

Figure 4.32 Air flapper valve assembly (100's only). (Courtesy of Lennox Industries, Inc.)

g. When placing a new or old valve back into the unit, line up the gasket (if used) and start all eight screws in place by hand (see note under item above).

Make sure that the star stamped on the cover plate or the words THIS SIDE OUT are placed outside of the unit. Then tighten the screws evenly to a maximum of 23 ± 1 oz in. for 50 and 80 series units and 10 ± 2 oz in. for 100 series units. Do not overtighten the screws. If the threads are damaged the entire valve body will have to be replaced. All eight screws must be in place for proper operation.

Checking the Gas Flapper Valve

Use the following steps when checking the gas flapper valve:

a. Disconnect the union at the bottom of the expansion tank and remove the entire flapper valve, nipple, and elbow assembly in one piece. It is not recommended to remove the elbow and nipple from the flapper valve unless the valve is being replaced. Use care not to damage the O-ring when handling the valve out of the unit. *Do not drop.*

b. Do not turn or remove the center screw of the valve assembly. Visually inspect the flapper. If the flapper is torn, creased, or has uneven (frayed) edges, the valve assembly must be replaced.

c. Check for free movement of the flapper over the spacer. Use a feeler gauge blade to carefully move the flapper between the plates. Be sure that the flapper is not trapped between the spacer and either the back plate or the valve body. If the flapper does not move freely or is trapped under the spacer, the valve assembly must be replaced.

d. Check for the required clearance between the flapper and the valve body (see Figure 4.33). Use a feeler gauge, starting at small and working up to the clearance dimensions until the gauge is just about snug. *Be very careful not to damage the flapper material by forcing the gauge.* The clearance should be checked around the valve in several places. If the valve is out

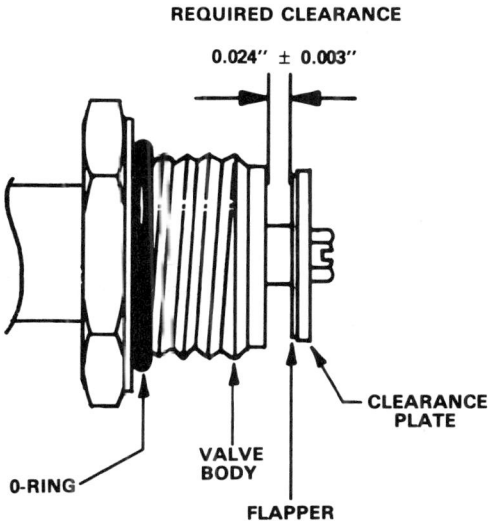

Figure 4.33 Gas flapper valve required clearance. (Courtesy of Lennox Industries, Inc.)

of the required clearance dimension given in Figure 4.33 at any point around the valve, it must be replaced.

e. When placing a new or old valve into the unit, use care not to damage the O-ring. *Do not use pipe sealers on the flapper valve threads.*

Checking the Gas Orifice

Use the following steps when checking the gas orifice:

a. With the gas flapper valve assembly removed, use a flashlight to check for blockage of the orifice in the manifold. Use a one-half-in. shallow socket with an extension to remove the orifice.

b. Check the orifice drill size for the unit as given by Lennox Repair Parts listings. If the orifice is incorrect it must be replaced.

c. Refer to Figure 4.34 for the physical characteristics of the orifice. The orifice opening must not be chamfered. The orifice taper must be centered and not recessed. *The taper must be on both sides of the orifice.* If any defects are found (including nicks or scars) the orifice must be replaced.

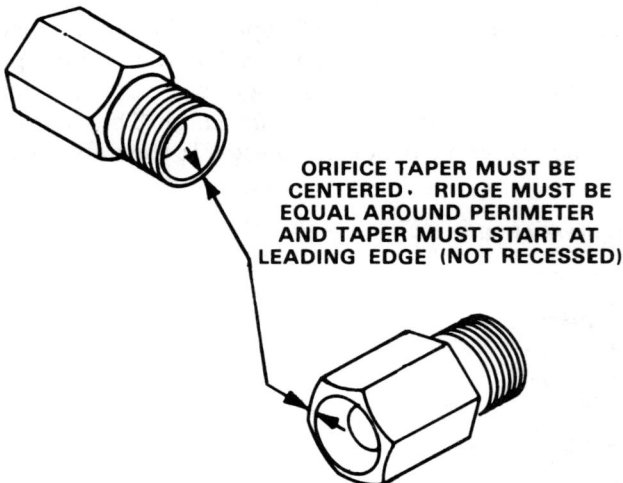

ORIFICE TAPER MUST BE
CENTERED. RIDGE MUST BE
EQUAL AROUND PERIMETER
AND TAPER MUST START AT
LEADING EDGE (NOT RECESSED)

ORIFICE MUST BE TAPERED ON BOTH SIDES

Figure 4.34 GSR14 series orifice characteristics. (Courtesy of Lennox Industries, Inc.)

d. Standard atmospheric burner orifices or orifice blanks cannot be used as replacements for the GSR14. Only replacement orifices supplied through Lennox should be used.

e. When threading the orifice into the manifold, use a rubber or foam insert in the socket to hold the threads past the end of the socket. *Carefully*

align the threads by hand, turning the socket and extension until the orifice is in place. Avoid crossthreading the orifice.

Other Problems

The following is a list of problems that might occur but are not covered under the four main troubleshooting categories. The steps for troubleshooting these problems are covered in the paragraphs following the list.

1. Blower runs continuously.
2. Frequent recycling of the blower after the heat cycle.
3. The supply air blower does not run.
4. The unit does not shut off.
5. Abnormal sounds.
6. Miscellaneous.

Troubleshooting other problems. Use the following steps when troubleshooting these types of problems:

1. Blower runs continuously:
   a. Is the room thermostat fan switch set to ON? If so, set to AUTO.
   b. Is the fan control OFF setting below the ambient air temperature? If so, readjust to 90 °F.
   c. Are the primary and auxiliary fan controls operating normally? Replace if necessary.
   d. Are the blower relay contacts operating normally? Replace if necessary.
2. Frequent recycling of the blower after the heat cycle:
   Check the primary fan control for correct adjustment. The primary fan control should be adjusted for 90 °F. Settings above 90 °F do not allow the heat exchanger assembly to cool down enough before the blower is stopped. The leftover heat retained in the unit when the airflow is stopped builds up, causing a fan recycle problem.
3. Supply air blower does not run:
   Be sure to reset the secondary limit if tripped.
   a. Check voltage at the blower tops after about 45 seconds.
   b. Check for loose wiring. Is the blower relay operating properly? Check for 120 VAC between terminal 2 and neutral immediately after ignition. See the electrical schematic.
   c. Is the fan control operating normally? If not, check for 120 VAC between terminal 5 and the neutral after the unit has been operating for 35 to 40 seconds. See the electrical schematic.

  d. Is the blower and/or capacitor operating normally? If not, check for 120 VAC between the heating motor tap and the neutral after the unit has been operating for 35 to 40 seconds. See the electrical schematic. Use standard motor troubleshooting techniques if the voltage reaches this point and the blower still does not operate.

4. Unit does not shut off:
  a. Is the room thermostat operating normally?
  b. Is the 24-VAC control wiring shorted? Check and repair.
  c. Is the gas valve stuck open?

5. Abnormal sounds:
     Abnormal hissing sounds around the air decoupler box may indicate one of the following. Corrective action is required if this sound is heard:
  a. Air leakage around the air decoupler box cover.
  b. Air leakage around the air decoupler box cover mounting screws.
  c. Air leakage out of the purge blower lead strain relief.
  d. Air leakage around the intake air connection to the air decoupler box.
  e. Air leakage out of the back of the air decoupler box:
    1. Around the intake pipe.
    2. Around the air decoupler box rear mounting bolts.

Danger: *Extremely loud pulse sounds, which can be easily heard through the supply or return air ducts, may indicate a combustion chamber or tailpipe leak. Do not allow the unit to operate with a combustion chamber or exhaust leak. Before the unit is allowed to operate, exhaust leaks must be located and corrected.*

The unit must be examined visually for unusual amounts of condensate in any area other than the condensing coil outlet which might indicate a system leak.

     Abnormal rattling and casing vibration other than obvious loose parts may indicate metal-to-metal contact of components that are normally separated during operation. The gas piping, condensing coil outlet, and the air intake pipe areas should be checked. The combustion chamber to exhaust decoupler relationship should also be checked.

6. Miscellaneous:
     To check for faulty auxiliary fan control, disconnect the control from the unit, and test for continuity. Contacts 1 and 3 should be open at ambient temperature and contacts 1 and 2 should be closed at ambient temperature.

> **Warning:** *Shock hazard. The auxiliary fan control is connected to line voltage. It is housed under a barrier for safety. Before servicing the control, be sure to disconnect the electric power to the unit.*

> **Warning:** *The barrier that shrouds the auxiliary fan control must be in place any time the unit is operating. Do not operate the unit without the barrier in place.*

To check for a faulty limit, disconnect the limit from the unit and test for continuity. The contacts should be closed at ambient temperature.

> **Note:** If the secondary limit will not reset after cooling several minutes, it is faulty and should be replaced.

LENNOX PULSE™ GSR14 GAS FURNACE UNIT WIRING DIAGRAM

GSR14 TROUBLESHOOTING FLOW CHART

NOTE - REFER TO ILLUSTRATIONS ON ADJACENT PAGE FOR NUMBERED CHECKS

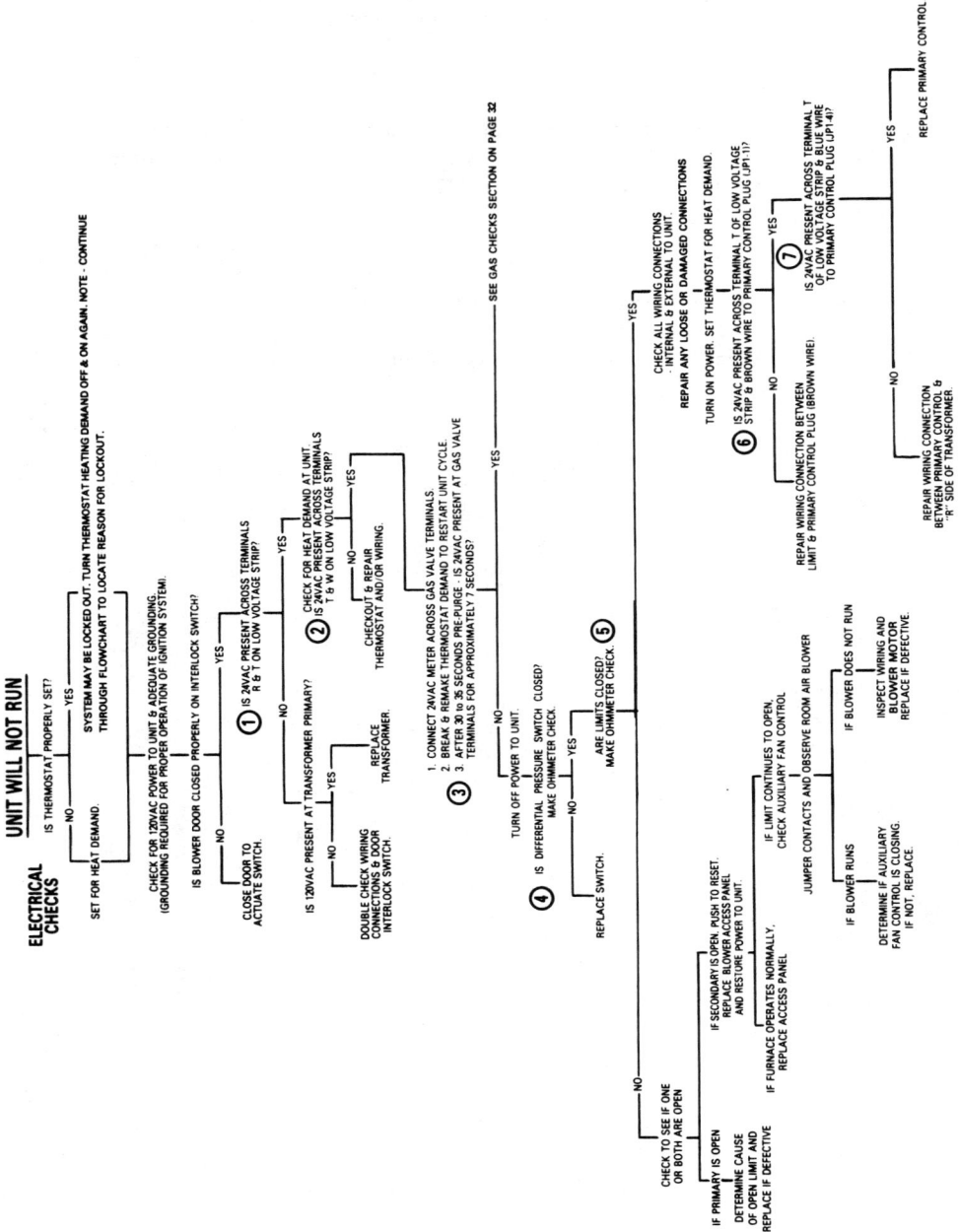

UNIT WILL NOT RUN

ELECTRICAL CHECKS

IS THERMOSTAT PROPERLY SET?

— NO → SET FOR HEAT DEMAND.

— YES → SYSTEM MAY BE LOCKED OUT. TURN THERMOSTAT HEATING DEMAND OFF & ON AGAIN. NOTE - CONTINUE THROUGH FLOWCHART TO LOCATE REASON FOR LOCKOUT.

CHECK FOR 120VAC POWER TO UNIT & ADEQUATE GROUNDING. (GROUNDING REQUIRED FOR PROPER OPERATION OF IGNITION SYSTEM)

IS BLOWER DOOR CLOSED PROPERLY ON INTERLOCK SWITCH?

— NO → CLOSE DOOR TO ACTUATE SWITCH.

— YES → ① IS 24VAC PRESENT ACROSS TERMINALS R & T ON LOW VOLTAGE STRIP?

IS 120VAC PRESENT AT TRANSFORMER PRIMARY?

— NO → DOUBLE CHECK WIRING CONNECTIONS & DOOR INTERLOCK SWITCH.

— YES → REPLACE TRANSFORMER.

② IS 24VAC PRESENT ACROSS TERMINALS T & W ON LOW VOLTAGE STRIP?

CHECK FOR HEAT DEMAND AT UNIT.

— NO → CHECKOUT & REPAIR THERMOSTAT AND/OR WIRING.

— YES →

③
1. CONNECT 24VAC METER ACROSS GAS VALVE TERMINALS.
2. BREAK & REMAKE THERMOSTAT DEMAND TO RESTART UNIT CYCLE.
3. AFTER 30 to 35 SECONDS PRE-PURGE - IS 24VAC PRESENT AT GAS VALVE TERMINALS FOR APPROXIMATELY 7 SECONDS?

— YES → SEE GAS CHECKS SECTION ON PAGE 32

④ TURN OFF POWER TO UNIT.

IS DIFFERENTIAL PRESSURE SWITCH CLOSED? MAKE OHMMETER CHECK.

— NO → REPLACE SWITCH.

— YES → ⑤ ARE LIMITS CLOSED? MAKE OHMMETER CHECK.

— NO → CHECK TO SEE IF ONE OR BOTH ARE OPEN

IF PRIMARY IS OPEN
DETERMINE CAUSE OF OPEN LIMIT AND REPLACE IF DEFECTIVE

IF SECONDARY IS OPEN, PUSH TO RESET. REPLACE BLOWER ACCESS PANEL AND RESTORE POWER TO UNIT.

IF LIMIT CONTINUES TO OPEN, CHECK AUXILIARY FAN CONTROL

IF FURNACE OPERATES NORMALLY, REPLACE ACCESS PANEL

JUMPER CONTACTS AND OBSERVE ROOM AIR BLOWER

IF BLOWER RUNS
DETERMINE IF AUXILIARY FAN CONTROL IS CLOSING. IF NOT, REPLACE.

IF BLOWER DOES NOT RUN
INSPECT WIRING AND BLOWER MOTOR REPLACE IF DEFECTIVE.

— YES → CHECK ALL WIRING CONNECTIONS - INTERNAL & EXTERNAL TO UNIT. REPAIR ANY LOOSE OR DAMAGED CONNECTIONS

TURN ON POWER. SET THERMOSTAT FOR HEAT DEMAND.

⑥ IS 24VAC PRESENT ACROSS TERMINAL T OF LOW VOLTAGE STRIP & BROWN WIRE TO PRIMARY CONTROL PLUG (JP1-1)?

— NO → REPAIR WIRING CONNECTION BETWEEN LIMIT & PRIMARY CONTROL PLUG (BROWN WIRE).

— YES → ⑦ IS 24VAC PRESENT ACROSS TERMINAL T OF LOW VOLTAGE STRIP & BLUE WIRE TO PRIMARY CONTROL PLUG (JP1-4)?

— NO → REPAIR WIRING CONNECTION BETWEEN PRIMARY CONTROL & "R" SIDE OF TRANSFORMER

— YES → REPLACE PRIMARY CONTROL.

GSR14 TROUBLESHOOTING — UNIT WILL NOT RUN

NOTE: CHECKS ILLUSTRATED CORRESPOND TO GSR14 TROUBLESHOOTING FLOW CHART ON OPPOSITE PAGE

163

GSR14 TROUBLESHOOTING FLOW CHART

NOTE - REFER TO ILLUSTRATIONS ON PREVIOUS PAGE FOR NUMBERED CHECKS.

CONTINUED FROM
UNIT-WILL-NOT-RUN
FLOWCHART ON
PAGE 30

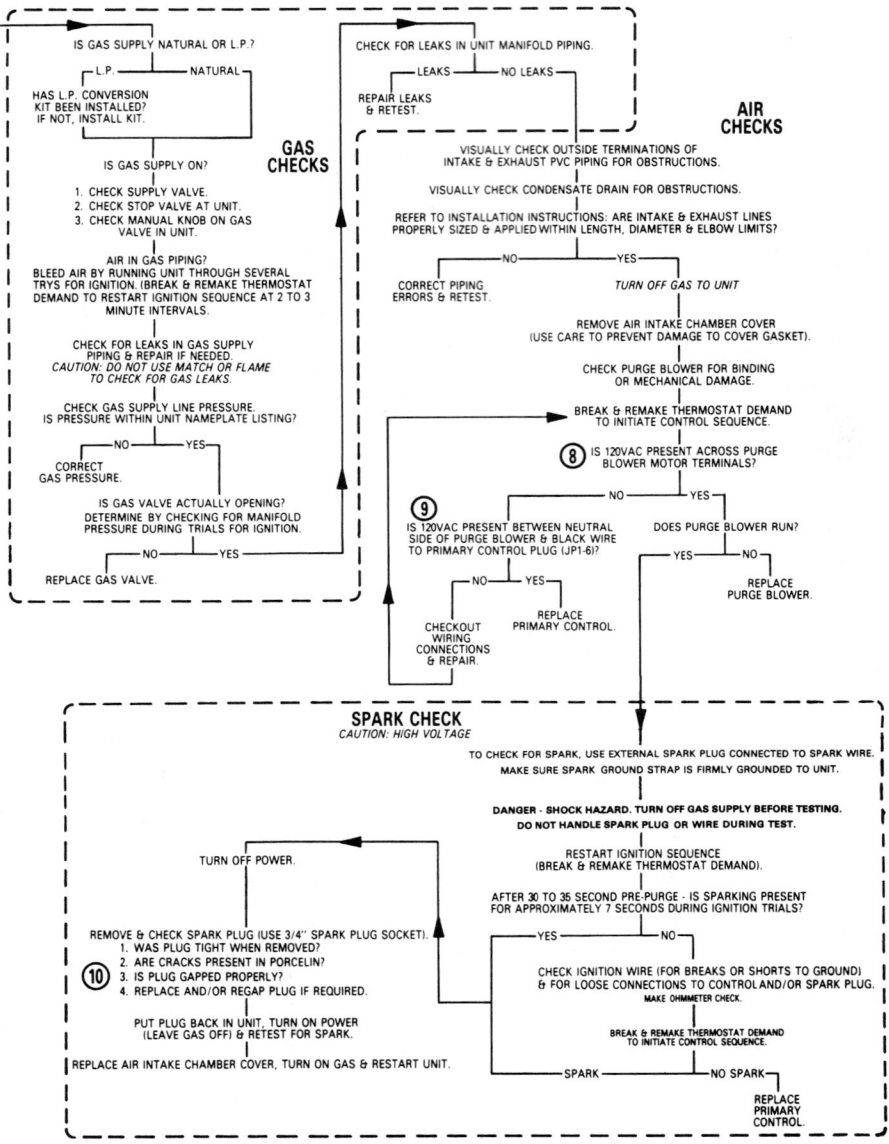

— YES

IS GAS SUPPLY NATURAL OR L.P.?

CHECK FOR LEAKS IN UNIT MANIFOLD PIPING.

— L.P. — NATURAL —

— LEAKS — NO LEAKS —

HAS L.P. CONVERSION
KIT BEEN INSTALLED?
IF NOT, INSTALL KIT.

REPAIR LEAKS
& RETEST.

**AIR
CHECKS**

**GAS
CHECKS**

IS GAS SUPPLY ON?

1. CHECK SUPPLY VALVE.
2. CHECK STOP VALVE AT UNIT.
3. CHECK MANUAL KNOB ON GAS
   VALVE IN UNIT.

VISUALLY CHECK OUTSIDE TERMINATIONS OF
INTAKE & EXHAUST PVC PIPING FOR OBSTRUCTIONS.

VISUALLY CHECK CONDENSATE DRAIN FOR OBSTRUCTIONS.

REFER TO INSTALLATION INSTRUCTIONS: ARE INTAKE & EXHAUST LINES
PROPERLY SIZED & APPLIED WITHIN LENGTH, DIAMETER & ELBOW LIMITS?

AIR IN GAS PIPING?
BLEED AIR BY RUNNING UNIT THROUGH SEVERAL
TRYS FOR IGNITION. (BREAK & REMAKE THERMOSTAT
DEMAND TO RESTART IGNITION SEQUENCE AT 2 TO 3
MINUTE INTERVALS.

— NO —— YES —

CORRECT PIPING
ERRORS & RETEST.

TURN OFF GAS TO UNIT

CHECK FOR LEAKS IN GAS SUPPLY
PIPING & REPAIR IF NEEDED.
*CAUTION: DO NOT USE MATCH OR FLAME
TO CHECK FOR GAS LEAKS.*

REMOVE AIR INTAKE CHAMBER COVER
(USE CARE TO PREVENT DAMAGE TO COVER GASKET).

CHECK GAS SUPPLY LINE PRESSURE.
IS PRESSURE WITHIN UNIT NAMEPLATE LISTING?

CHECK PURGE BLOWER FOR BINDING
OR MECHANICAL DAMAGE.

— NO —— YES —

BREAK & REMAKE THERMOSTAT DEMAND
TO INITIATE CONTROL SEQUENCE.

CORRECT
GAS PRESSURE.

(8) IS 120VAC PRESENT ACROSS PURGE
     BLOWER MOTOR TERMINALS?

IS GAS VALVE ACTUALLY OPENING?
DETERMINE BY CHECKING FOR MANIFOLD
PRESSURE DURING TRIALS FOR IGNITION.

— NO —

— YES —

(9)

DOES PURGE BLOWER RUN?

— NO —— YES —

IS 120VAC PRESENT BETWEEN NEUTRAL
SIDE OF PURGE BLOWER & BLACK WIRE
TO PRIMARY CONTROL PLUG (JP1-6)?

— YES —— NO —

REPLACE GAS VALVE.

REPLACE
PURGE BLOWER.

— NO —— YES —

CHECKOUT
WIRING
CONNECTIONS
& REPAIR.

REPLACE
PRIMARY CONTROL.

SPARK CHECK
CAUTION: HIGH VOLTAGE

TO CHECK FOR SPARK, USE EXTERNAL SPARK PLUG CONNECTED TO SPARK WIRE.
MAKE SURE SPARK GROUND STRAP IS FIRMLY GROUNDED TO UNIT.

**DANGER - SHOCK HAZARD. TURN OFF GAS SUPPLY BEFORE TESTING.
DO NOT HANDLE SPARK PLUG OR WIRE DURING TEST.**

TURN OFF POWER.

RESTART IGNITION SEQUENCE
(BREAK & REMAKE THERMOSTAT DEMAND).

AFTER 30 TO 35 SECOND PRE-PURGE - IS SPARKING PRESENT
FOR APPROXIMATELY 7 SECONDS DURING IGNITION TRIALS?

— YES —— NO —

REMOVE & CHECK SPARK PLUG (USE 3/4" SPARK PLUG SOCKET).
1. WAS PLUG TIGHT WHEN REMOVED?
2. ARE CRACKS PRESENT IN PORCELIN?
(10) 3. IS PLUG GAPPED PROPERLY?
4. REPLACE AND/OR REGAP PLUG IF REQUIRED.

CHECK IGNITION WIRE (FOR BREAKS OR SHORTS TO GROUND)
& FOR LOOSE CONNECTIONS TO CONTROL AND/OR SPARK PLUG.
MAKE OHMMETER CHECK.

PUT PLUG BACK IN UNIT, TURN ON POWER
(LEAVE GAS OFF) & RETEST FOR SPARK.

BREAK & REMAKE THERMOSTAT DEMAND
TO INITIATE CONTROL SEQUENCE.

REPLACE AIR INTAKE CHAMBER COVER, TURN ON GAS & RESTART UNIT.

— SPARK —— NO SPARK —

REPLACE
PRIMARY
CONTROL.

UNIT SPUTTER STARTS & DIES

GAS CHECKS

IS GAS SUPPLY NATURAL OR L.P.?

L.P. — NATURAL

HAS L.P. CONVERSION KIT BEEN INSTALLED? IF NOT, INSTALL KIT.

IS GAS SUPPLY ON?

1. CHECK SUPPLY VALVE.
2. CHECK STOP VALVE AT UNIT
3. CHECK MANUAL KNOB ON GAS VALVE IN UNIT.

AIR IN GAS PIPING? BLEED AIR BY RUNNING UNIT THROUGH SEVERAL TRYS FOR IGNITION. (BREAK & REMAKE THERMOSTAT DEMAND TO RESTART IGNITION SEQUENCE AT 2 TO 3 MINUTE INTERVALS.

CHECK FOR LEAKS IN GAS SUPPLY PIPING & REPAIR IF NEEDED. *CAUTION: DO NOT USE MATCH OR FLAME TO CHECK FOR GAS LEAKS.*

CHECK GAS SUPPLY LINE PRESSURE. IS PRESSURE WITHIN UNIT NAMEPLATE LISTING?

NO — YES

CORRECT GAS PRESSURE.

IS MANIFOLD PRESSURE CORRECT?

NO — YES

ADJUST GAS VALVE REGULATOR FOR PROPER MANIFOLD PRESSURE

AIR CHECKS

CHECK FOR LEAKS IN UNIT MANIFOLD PIPING.

LEAKS — NO LEAKS

REPAIR LEAKS & RETEST.

VISUALLY CHECK OUTSIDE TERMINATIONS OF INTAKE & EXHAUST PVC PIPING FOR OBSTRUCTIONS.

VISUALLY CHECK CONDENSATE DRAIN FOR OBSTRUCTIONS.

REFER TO INSTALLATION INSTRUCTIONS: ARE INTAKE & EXHAUST LINES PROPERLY SIZED & APPLIED WITHIN LENGTH, DIAMETER & ELBOW LIMITS?

NO — YES

CORRECT PIPING ERRORS & RETEST.

CHECK FOR AIR LEAKS AROUND AIR INTAKE CHAMBER COVER SEAL & SEALS AT BACK OF CHAMBER AROUND AIR MANIFOLD.

TURN OFF GAS TO UNIT

REMOVE AIR INTAKE CHAMBER COVER (USE CARE TO PREVENT DAMAGE TO COVER GASKET).

CHECK PURGE BLOWER FOR BINDING OR MECHANICAL DAMAGE.

BREAK & REMAKE THERMOSTAT DEMAND TO INITIATE CONTROL SEQUENCE.

REFER TO ILLUSTRATIONS ON 'UNIT WILL NOT RUN' FLOW CHART.

(8) IS 120VAC PRESENT ACROSS PURGE BLOWER MOTOR TERMINALS?

NO — YES

(9) IS 120VAC PRESENT BETWEEN NEUTRAL SIDE OF PURGE BLOWER & BLACK WIRE TO PRIMARY CONTROL PLUG (JP1-6)?

DOES PURGE BLOWER RUN?

YES — NO

REPLACE PURGE BLOWER.

NO — YES

CHECKOUT WIRING CONNECTIONS & REPAIR.

REPLACE PRIMARY CONTROL.

IS FLAME SIGNAL PRESENT & CORRECT? CHECK WITH MICROAMP METER.

NO — YES

CHECK SENSOR WIRE. REPLACE IF DETERIORATED & RETEST UNIT.

REMOVE & CHECK SENSOR PLUG (USE 11/16" SPARK PLUG SOCKET).

1. WAS PLUG TIGHT WHEN REMOVED?
2. ARE CRACKS PRESENT IN PORCELIN?
3. IS SENSOR ELECTRODE CORRODED OR DAMAGED? REPLACE SENSOR PLUG IF DEFECTIVE & RETEST UNIT.

REPLACE PRIMARY CONTROL IF DEFECTIVE.

IS DIFFERENTIAL PRESSURE SWITCH CUTTING OUT UNIT?

MAKE OHMMETER CHECK ACROSS TERMINALS IMMEDIATELY AFTER UNIT SHUTS OFF. METER READS ∞ OHMS WHEN SWITCH CUTS UNIT OUT.

NO — YES

IS AIR FLAPPER VALVE OPERATING NORMALLY (LACK OF AIR)? REFER TO "CHECKING AIR INTAKE INTAKE FLAPPER VALVE"

YES — NO

REPLACE AIR FLAPPER VALVE.

IS GAS FLAPPER VALVE OPERATING NORMALLY? REFER TO "CHECKING GAS FLAPPER VALVE"

YES — NO

REPLACE GAS FLAPPER VALVE.

CHECK FOR PARTIAL BLOCKAGE OF ALL PVC PIPING & CONDENSATE LINE.

IF NO BLOCKAGE EXISTS - REPLACE DIFFERENTIAL PRESSURE SWITCH

IS GAS ORIFICE CORRECT? REFER TO "CHECKING GAS ORIFICE"

REPLACE ORIFICE IF NECESSARY.

UNIT STARTS CLEAN BUT RUNS LESS THAN 10 SECONDS

RESET UNIT IF LOCKED OUT. LISTEN FOR CHANGE IN SOUND OF UNIT BEFORE IT STOPS.

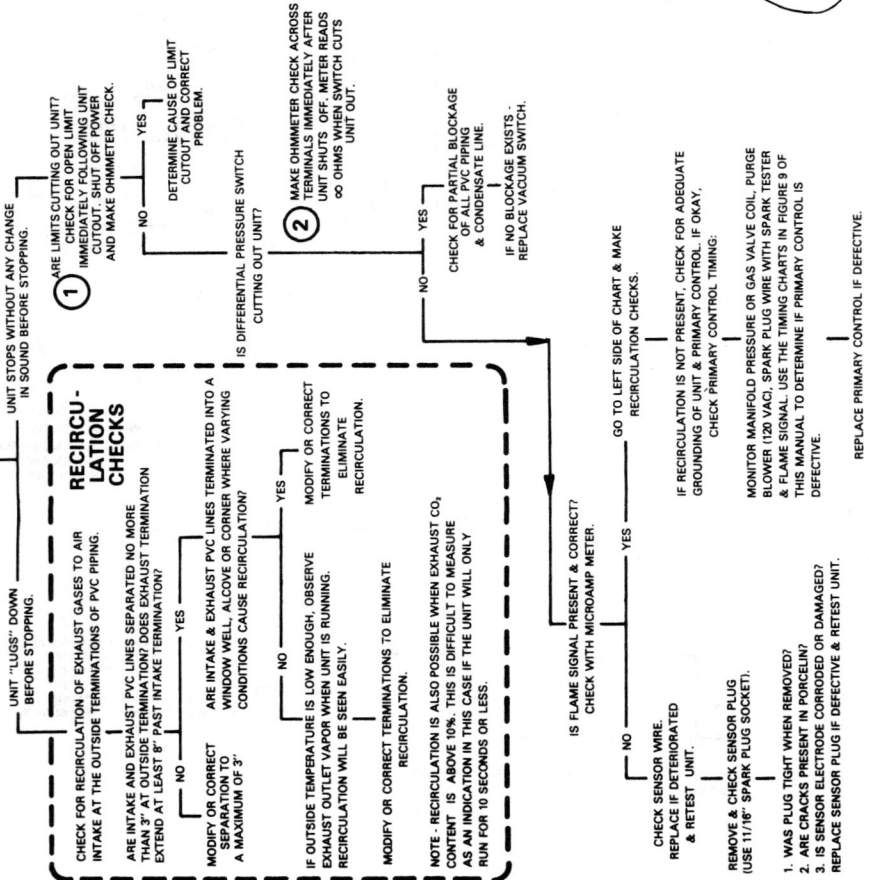

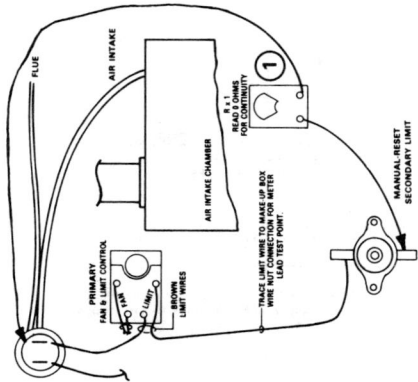

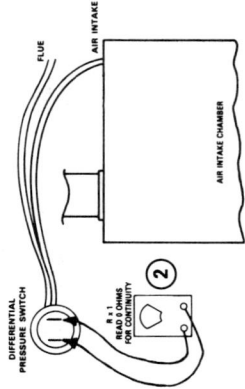

UNIT "LUGS" DOWN BEFORE STOPPING.

UNIT STOPS WITHOUT ANY CHANGE IN SOUND BEFORE STOPPING.

RECIRCU-LATION CHECKS

CHECK FOR RECIRCULATION OF EXHAUST GASES TO AIR INTAKE AT THE OUTSIDE TERMINATIONS OF PVC PIPING.

ARE INTAKE AND EXHAUST PVC LINES SEPARATED NO MORE THAN 3" AT OUTSIDE TERMINATION? DOES EXHAUST TERMINATION EXTEND AT LEAST 6" PAST INTAKE TERMINATION?

NO — MODIFY OR CORRECT SEPARATION TO A MAXIMUM OF 3"

YES — ARE INTAKE & EXHAUST PVC LINES TERMINATED INTO A WINDOW WELL, ALCOVE OR CORNER WHERE VARYING CONDITIONS CAUSE RECIRCULATION?

YES — MODIFY OR CORRECT TERMINATIONS TO ELIMINATE RECIRCULATION.

NO — IF OUTSIDE TEMPERATURE IS LOW ENOUGH, OBSERVE EXHAUST OUTLET VAPOR WHEN UNIT IS RUNNING. RECIRCULATION WILL BE SEEN EASILY.

MODIFY OR CORRECT TERMINATIONS TO ELIMINATE RECIRCULATION.

NOTE - RECIRCULATION IS ALSO POSSIBLE WHEN EXHAUST CO_2 CONTENT IS ABOVE 10%. THIS IS DIFFICULT TO MEASURE AS AN INDICATION IN THIS CASE IF THE UNIT WILL ONLY RUN FOR 10 SECONDS OR LESS.

---

(1) ARE LIMITS CUTTING OUT UNIT? CHECK FOR OPEN LIMIT IMMEDIATELY FOLLOWING UNIT CUTOUT. SHUT OFF POWER AND MAKE OHMMETER CHECK.

NO

YES — DETERMINE CAUSE OF LIMIT CUTOUT AND CORRECT PROBLEM.

IS DIFFERENTIAL PRESSURE SWITCH CUTTING OUT UNIT?

(2) MAKE OHMMETER CHECK ACROSS TERMINALS IMMEDIATELY AFTER UNIT SHUTS OFF. METER READS ∞ OHMS WHEN SWITCH CUTS UNIT OUT.

YES — CHECK FOR PARTIAL BLOCKAGE OF ALL PVC PIPING & CONDENSATE LINE.

IF NO BLOCKAGE EXISTS - REPLACE VACUUM SWITCH.

NO

GO TO LEFT SIDE OF CHART & MAKE RECIRCULATION CHECKS.

IS FLAME SIGNAL PRESENT & CORRECT? CHECK WITH MICROAMP METER.

NO — CHECK SENSOR WIRE. REPLACE IF DETERIORATED & RETEST UNIT.

REMOVE & CHECK SENSOR PLUG (USE 11/16" SPARK PLUG SOCKET).
1. WAS PLUG TIGHT WHEN REMOVED?
2. ARE CRACKS PRESENT IN PORCELIN?
3. IS SENSOR ELECTRODE CORRODED OR DAMAGED?
REPLACE SENSOR PLUG IF DEFECTIVE & RETEST UNIT.

YES — IF RECIRCULATION IS NOT PRESENT, CHECK FOR ADEQUATE GROUNDING OF UNIT & PRIMARY CONTROL. IF OKAY, CHECK PRIMARY CONTROL TIMING:

MONITOR MANIFOLD PRESSURE OR GAS VALVE COIL, PURGE BLOWER (120 VAC), SPARK PLUG WIRE WITH SPARK TESTER & FLAME SIGNAL. USE THE TIMING CHARTS IN FIGURE 9 OF THIS MANUAL TO DETERMINE IF PRIMARY CONTROL IS DEFECTIVE.

REPLACE PRIMARY CONTROL IF DEFECTIVE.

FLUE
AIR INTAKE
AIR INTAKE CHAMBER
(1)
R x 1 READ 0 OHMS FOR CONTINUITY
MANUAL-RESET SECONDARY LIMIT
PRIMARY FAN & LIMIT CONTROL
BROWN LIMIT WIRES
TRACE LIMIT WIRE TO MAKE UP BOX WIRE NUT CONNECTION FOR METER LEAD TEST POINT.

FLUE
AIR INTAKE
AIR INTAKE CHAMBER
(2)
R x 1 READ 0 OHMS FOR CONTINUITY
DIFFERENTIAL PRESSURE SWITCH

UNIT RUNS BUT SHUTS OFF BEFORE THERMOSTAT IS SATISFIED - INSUFFICIENT HEAT

(1) DOES EITHER LIMIT CUTOUT CAUSING UNIT TO SHUT OFF? TURN OFF POWER & MAKE OHMMETER CHECK IMMEDIATELY FOLLOWING CUTOUT OF UNIT.

NO

IS GAS PRESSURE LOW? CHECK FOR INTERMITTENT LOW GAS PRESSURE. (MONITOR MANIFOLD PRESSURE)

YES — CORRECT GAS PRESSURE.

NO — IS EXHAUST, INTAKE OR CONDENSATE PVC PIPING PARTIALLY BLOCKED OR RESTRICTED?

YES — ELIMINATE BLOCKAGE & RETEST.

NO →

IS FILTER CLEAN & PROPERLY INSTALLED? CORRECT IF NECESSARY & RETEST UNIT.

ARE SUPPLY & RETURN AIR DUCTS UNRESTRICTED? CORRECT IF NECESSARY & RETEST.

CHECK GAS BTUH INPUT.

IS INPUT HIGH? OUT OF ACCEPTABLE RANGE?

YES — IS MANIFOLD RUNNING PRESSURE TOO HIGH?

YES — ADJUST GAS VALVE REGULATOR

NO →

NO — IS TEMPERATURE RISE AND STATIC PRESSURE WITHIN PROPER RANGE?

YES — REPLACE LIMIT.

NO — ADJUST BLOWER SPEED.

RECIRCULATION CHECKS

CHECK FOR RECIRCULATION OF EXHAUST GASES TO AIR INTAKE AT THE OUTSIDE TERMINATION OF PVC EXHAUST PIPING.

ARE INTAKE & EXHAUST PVC LINES SEPARATED NO MORE THAN 3" AT OUTSIDE TERMINATION? DOES EXHAUST TERMINATION EXTEND AT LEAST 8" PAST INTAKE TERMINATION?

NO — MODIFY OR CORRECT SEPARATION TO A MAXIMUM OF 3".

YES — ARE INTAKE & EXHAUST PVC LINES TERMINATED INTO A WINDOW WELL, ALCOVE OR CORNER WHERE VARYING CONDITIONS CAUSE RECIRCULATION?

NO →

YES — IF OUTSIDE TEMPERATURE IS LOW ENOUGH, OBSERVE EXHAUST OUTLET VAPOR WHEN UNIT IS RUNNING. RECIRCULATION WILL BE SEEN EASILY.

IF OUTSIDE TEMPERATURE IS TOO HIGH TO SEE EXHAUST VAPOR OR RECIRCULATION CANNOT BE DETERMINED. CHECK CO_2 CONTENT OF EXHAUST GAS. IF CO_2 CONTENT IS ABOVE 10% RECIRCULATION IS POSSIBLE.

FLUE

AIR INTAKE

AIR INTAKE CHAMBER

R x 1 READ 0 OHMS FOR CONTINUITY

(1)

MANUAL RESET SECONDARY LIMIT

PRIMARY FAN & LIMIT CONTROL

BROWN LIMIT WIRES

TRACE LIMIT WIRE TO MAKE UP BOX WIRE NUT CONNECTION FOR WHITE LEAD TEST POINT

5

Coleman T.H.E. 90 Gas Furnace ("___" and "A" Models Only)

These instructions are intended for the use of qualified individuals specially trained and experienced in servicing this type of equipment and related system components. Service personnel are required by some states to be licensed. Persons not qualified should not attempt to install or service this equipment. Improper installation or service may damage the equipment, will void the warranty, and may create a hazard.

This is not a basic heating manual and does not, therefore, cover the basic principles of heating. The user of this manual should have already accomplished a thorough study of heating and should use this manual as an advanced text to apply only to the Coleman furnace models listed in the specification table (see Table 5.1).

Being service oriented, this guide also does not cover all the details of the furnace installation. Installation instructions are packed with each furnace and copies are available upon request.

The furnaces covered in this guide are high-efficiency models, featuring direct spark ignition (no pilot) and a vent blower to provide a constant supply of combustion air to the burners.

The direct spark ignition system is designed to provide a "purge" cycle so that any unburned gas that might have accumulated in the heat exchanger is dissipated before ignition occurs.

TABLE 5.1   FURNACE SPECIFICATIONS (COURTESY OF THE COLEMAN COMPANY)

Model	Nat.	2940–666	2960–666	2970–766	2985–766	2960–766
Unit rating	Input: 0-2,000 ft. elevation	45,000	65,000	80,000	95,000	65,000
Btu/hr.	High altitude	For elevations above 2,000 ft. reduce input 4% for each 1,000 ft. of elev. above sea level				
Air temperature rise range °F		20-50	25-55	30-60	35-65	25-55
Designed max. outlet air temp. °F		145	150	155	160	150
Max. external static pressure in. W.C.		.5				
Vent pipe		2 in. diameter * schedule 40 CPVC				
Vent pipe fittings		2 in. diameter schedule 40 or 80 CPVC				
Condensate pipe and fittings		1/2 in. CPVC				
Gas connection		1/2 in. FPT				
Electrical service		115 VAC 60 Hz 1 Ph				
Minimum distance to combustible materials—inches Front Back Plenum-top Floor Sides Vent		6 0 1 Combustible 0 0				
Filter (furnished)		16 × 25 × 1/2		20 × 25 × 1/2		

*3 in. diameter must be used for all horizontal runs with model 2985.

An air pressure switch prevents furnace operation if for some reason inadequate combustion air is being provided.

SEQUENCE OF OPERATION

These furnaces are equipped with an electric spark, direct burner ignition system; therefore, in response to a call for heat by the room thermostat, the burner is lighted by an electric arc at the beginning of each operation cycle. The burner will continue to operate until the thermostat is satisfied, at which time all burner flame is extinguished. During the off cycle, no gas energy is consumed. With the room thermostat set below room temperature and with the electrical power and gas supply to the furnace on, the normal sequence of operation is as follows:

1. When the room temperature falls below the setting of the room thermostat, the thermostat energizes the heating relay.

2. When the heating relay closes, a circuit is made, starting the vent blower. A circuit is also made through the normally closed limit switch contacts and the heating relay to the normally open vent air pressure switch contacts.

3. As the blower increases in speed, a negative pressure is developed in the vent blower. When sufficient negative pressure has been developed, the contacts of the vent air pressure switch will close and complete the electrical circuit to the electronic ignition module.

4. After 15 to 20 seconds the electronic ignition module simultaneously energizes (1) the electric ignition electrode, (2) the gas control valve, (3) the blower sequencer, and (4) a safety lockout.

5. When the burner lights, a flame sensor, which is part of the ignition electrode assembly, senses the presence of the burner flame and causes the safety lockout circuit in the ignition module to be deenergized. This allows the gas valve to remain open and the burner to operate.

6. If the burner fails to light within 6 to 8 seconds from the time the ignition control is first energized by the vent pressure switch, the ignition module will deenergize (1) the gas valve, (2) the ignition electrode, and (3) the blower sequencer. The ignition control will again energize (1) the gas valve, (2) the ignition electrode, and (3) the blower sequencer after a 15- to 20-second delay. If after three trials for ignition the burner still fails to light, the safety lockout circuit in the ignition module will deenergize and lock off (1) the gas valve, (2) the ignition electrode, and (3) the blower sequencer. The system will remain in a lockout mode until the room thermostat is set below room temperature, causing the thermostat contacts to open and release the safety lockout circuit. Then setting the room thermostat above room temperature and causing the thermostat contacts to close will start the system to try for ignition again.

7. The lapsed time from the moment the wall thermostat closes to when the burner lights may be from 20 to 40 seconds. This delay in the ignition sequence is caused by (1) the time required for the vent blower to develop sufficient negative pressure to activate the vent air pressure switch, (2) the somewhat slow reaction of the vent air pressure switch, and (3) the 15- to 20-second delay designed into the ignition module. The lapsed time

will also be affected by the temperature within the furnace and the vent piping.

The 15- to 20-second delay designed into the ignition module is a purge cycle. This allows the vent blower time to replenish the heat exchanger with fresh air so ignition can occur safely.

8. Thirty to forty seconds after the burner has lighted, the normally open contacts of the blower sequencer close and the furnace air circulation blower runs. The fan switch, which is in a parallel circuit with the blower sequencer, will also turn on but at a later time.

> **Note:** If a heating/cooling thermostat is being used and the fan switch is set in the ON (continuous blower) position, the furnace air circulation blower will run at the air conditioning speed. If the wall thermostat calls for heat, the furnace air circulation blower will shut off for 20 to 40 seconds, then the burner will light. Thirty to fifty seconds after the burner has lighted, the furnace air circulation blower will begin running again, but at the heating speed. There is no pause in the furnace air circulation blower operation when the wall thermostat is satisfied; the furnace air circulation blower just changes over to cooling (continuous blower) speed.

9. When the room thermostat is satisfied, the circuit to the heating relay is broken and the heating relay contacts return to the normally open position. The circuit to the vent blower, the ignition module, and the blower sequencer is broken and the burner is extinguished. The contacts of the blower sequencer return to the normally open position within 30 seconds after the burner extinguishes. The fan switch remains on until it senses a fall in the heat exchanger temperature to a safe limit then opens the circuit to the furnace air circulation blower.

PREOPERATIONAL CHECKS

> **Danger:** *Shock hazard:* Be sure that the electrical power to the furnace is turned off before performing the preoperational checks.

The following is a list of the preoperational checks that should be performed before operating the furnace:

1. Be sure that the furnace is equipped for the type of gas being supplied to the furnace. See the unit rating plate.

2. Make sure that the shipping strap was removed from the blower housing.

3. Manually spin the circulating air blower wheel to ensure that it turns freely and does not strike the blower housing.

4. Was the vent blower checked to ensure that it turned freely before the vent pipe was attached? See the venting instructions in Chapter 1.

5. Was the gas piping pressure tested and/or purged of air then checked for leaks? See gas piping in Chapter 1. Even the smallest leak must be eliminated before attempting to light the furnace.

OPERATION CHECKS

The gas valve is multifunctional, with two operating valves in line, pressure regulator, and manual gas cock (see Figures 5.1–5.3).

The pressure regulator is factory set to provide an operating manifold gas pressure of 3.5 in. wc on models equipped for natural gas and 10 in. wc on models equipped for LP gas.

The pressure regulator is a pressure limited adjustment type. Refer to the section Minor Input Adjustment under Furnace Input Capacity below.

Lighting Instructions

This furnace is equipped with an automatic electric spark direct burner ignition system that lights the main burner each time the thermostat calls for heat.

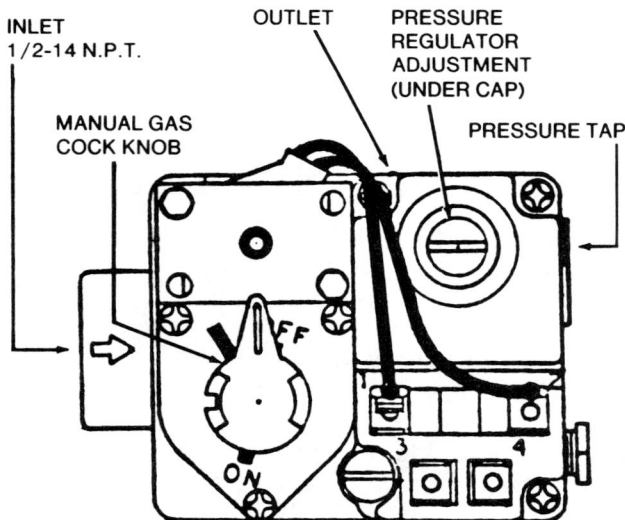

Figure 5.1 White Rodgers gas valve. (Courtesy of The Coleman Company.)

INLET
1/2-14 N.P.T.

OUTLET

PRESSURE
REGULATOR
ADJUSTMENT
(UNDER CAP)

MANUAL GAS
COCK KNOB

PRESSURE TAP

PUSH-TURN ON

OFF

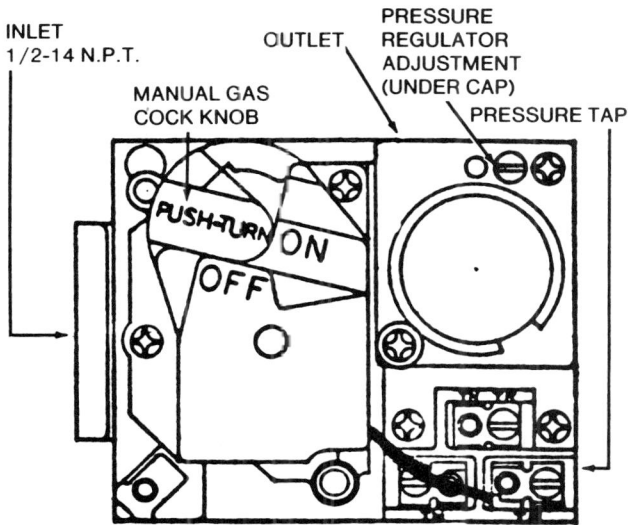

Figure 5.2 Honeywell gas valve.
(Courtesy of The Coleman Company.)

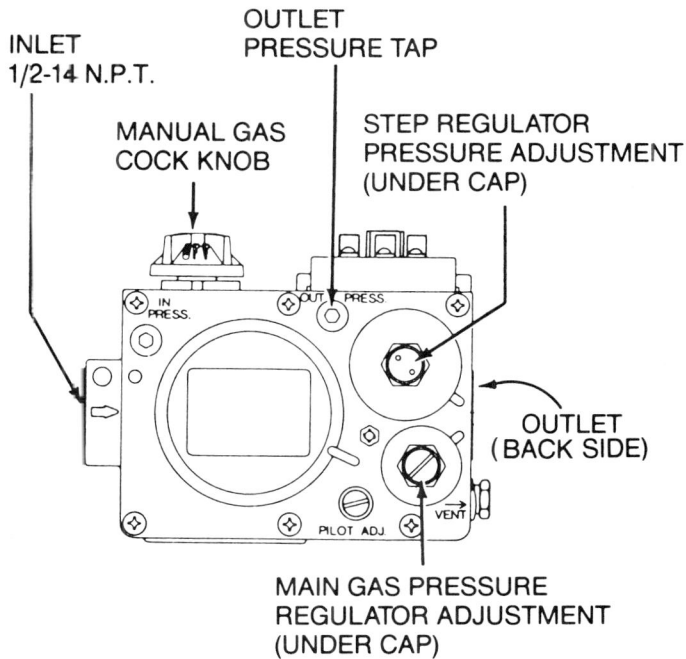

INLET
1/2-14 N.P.T.

OUTLET
PRESSURE TAP

MANUAL GAS
COCK KNOB

STEP REGULATOR
PRESSURE ADJUSTMENT
(UNDER CAP)

IN PRESS.

OUT. PRESS.

OUTLET
(BACK SIDE)

PILOT ADJ.

VENT

MAIN GAS PRESSURE
REGULATOR ADJUSTMENT
(UNDER CAP)

Figure 5.3 Robertshaw gas valve. (Courtesy of The Coleman Company.)

This furnace cannot be lighted with a match.

Warning: Failure to follow these instructions may result in an explosion
and possible damage to the furnace and injury to the operator.

Use the following steps to light this furnace:

1. Set the room thermostat to the lowest setting or OFF.
2. Turn the knob on the gas control valve to OFF. The White Rodgers gas
   valve has a built-in stop that prevents the knob from being turned directly to OFF. At this stop the knob must be depressed and then turned
   to OFF.
3. Wait 5 minutes.
4. Turn the knob on the gas control to the ON position.
5. Set the room thermostat to the desired setting, above room temperature;
   the burner will light, which may take 20 to 40 seconds.
6. If after three trials for ignition, the burner fails to light, go to the complete shutdown mode and determine the cause for failure for the furnace
   to light. Be sure that the gas supply to the furnace is turned on and the
   supply piping has been purged of air. Be sure that the electric power to
   the furnace is on.
7. For a complete shutdown, set the room thermostat to the lowest setting
   or to OFF and turn the knob on the gas control valve to OFF.

Burner Adjustment

After lighting the furnace, allow it to operate for approximately 15 minutes
and then adjust the burner primary air as follows:

1. Loosen the air adjustment rod locking screw (see Figure 5.4).
2. Close the primary air shutter by pulling on the adjustment rod until yellow tips appear in the flame at the end of the burner (see Figure 5.5).
3. Slowly open the primary air shutter by pushing the adjustment rod in
   until the yellow tips at the end of the burner disappear; then push the
   adjustment rod in another $\frac{1}{8}$ in.
4. Secure the primary air adjustment rod by tightening the locking screw
   against it.

VALVE BRACKET
ATTACHMENT
SCREWS (TWO
EACH SIDE)

LOCKING SCREW

PRIMARY AIR
ADJUSTMENT ROD

Figure 5.4 Burner primary air adjustment. (Courtesy of The Coleman Company.)

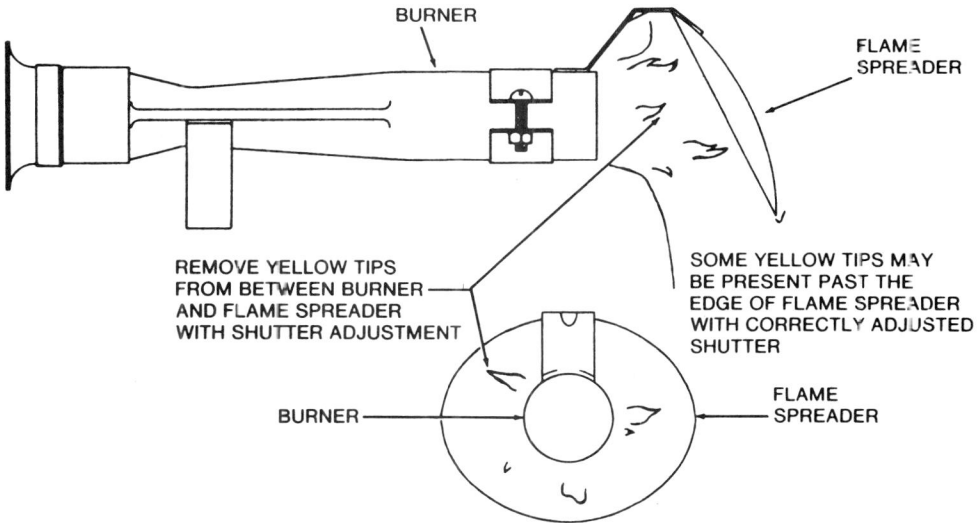

BURNER

FLAME
SPREADER

REMOVE YELLOW TIPS
FROM BETWEEN BURNER
AND FLAME SPREADER
WITH SHUTTER ADJUSTMENT

SOME YELLOW TIPS MAY
BE PRESENT PAST THE
EDGE OF FLAME SPREADER
WITH CORRECTLY ADJUSTED
SHUTTER

BURNER

FLAME
SPREADER

Figure 5.5 Flame adjustment. (Courtesy of The Coleman Company.)

Furnace Input Capacity

The maximum Btuh input capacity for each model is shown on the unit rating plate and in the specification tables (Table 5.1). This input must not be exceeded.

The input shown may be used in geographic areas where the elevation is from 0 to 2000 ft above sea level. In areas above 2000 ft (high altitude) the furnace Btu input must be reduced 4% for each 1000 ft of elevation above sea level. The Btu input depends on the calorific value of the gas (Btu/ft^3), orifice size, and manifold pressure. Coleman orifice sizes are based on calorific value of 1050 Btu/ft^3 for natural gas and 2500 Btu/ft^3 for LP gas (propane). The orifice size supplied with the furnace should provide satisfactory input capacity for installations in most areas, except at high altitude.

Minor input adjustment.   The input may be adjusted slightly by adjusting the pressure regulator in the gas control valve to change the manifold pressure.

To adjust the pressure regulator, remove the cover screw (see the locations in Figures 5.1–5.3) on top of the valve. Turn the adjusting screw counterclockwise to decrease the pressure, and turn it clockwise to increase the pressure. In no case should the final manifold pressure vary more than plus or minus 0.3 in. wc from the specified regulator pressure settings—3.5 in. wc for natural gas and 10 in. wc for LP gas.

Danger: Never attempt to modify this furnace: fire, explosion, or asphyxiation may result.

Determining gas input rate.   Where the gas is metered, the input rate may be determined by the following method.

Contact the gas supplier, public utility company, or LP gas distributor to obtain the calorific value of the gas being used. When checking the input rate, any other gas-burning appliances connected to the same meter should be completely off. The furnace should be allowed to operate for approximately 15 minutes before attempting to check the gas flow rate.

To check the flow rate, observe the 1-ft^3 dial on the gas meter and determine the number of seconds required for the dial to make one revolution (seconds to flow 1 ft^3).

To determine the number of seconds required for the flow of 1 ft^3 of gas, use the following formula:

$$\text{calorific value (Btu content) of gas} \times 3600 \div \text{furnace Btuh input}$$

EXAMPLE

1000 Btu gas, furnace input 100 000 Btu/hour. Seconds for 1 ft^3 = 1000 × 3600 ÷ 100 000 = 36 seconds

If when clocking the meter, the 1-ft^3 dial makes a complete revolution in less time than was calculated that it should, the furnace is overfired and will need to be derated. If in clocking the meter, it takes more time for the meter to make one revolution, the furnace is underfired.

The orifice size must be changed to correct an overfired or underfired condition. If it is determined that a different orifice is needed, be sure to replace it with the correct replacement to eliminate the problem.

Balance the System

It is recommended that the air distribution system be balanced, using in-line duct dampers if employed, to provide for satisfactory air delivery room to room. It is recommended that dampers in registers not be used for balancing the system.

Cycle Furnace for Correct Operation

After all parts of the installation have been completed, furnace installation, vent installation, and duct system installation, the furnace should be checked for normal operation prior to the time that the system will be used. Use the following procedures for this operation:

1. Place the heating system in the operating mode by following the lighting instructions outlined previously in these instructions and on the furnace rating plate.
2. Cycle the furance off and on a number of times, allowing the furnace to light and the circulating air blower to come on. After the thermostat is turned off allow time for both circulating air blower and the vent blower to stop before the next operating cycle.
3. While the unit is in operation, check to determine that the circulating air blower is operating smoothly with no undue vibration or noise. Check to determine that the vent blower is operating smoothly without noise or vibration. Check to ensure that the main burner has been adjusted and that the main burner flame is normal.
4. Turn off the furnace at the thermostat and turn off the electric power supply to the furnace. Disconnect the black wire from the vent blower motor and isolate the wire terminal to prevent any contact with the furnace casing or ground to eliminate the possibility of a shock hazard. Turn on the electric power to the furnace and turn on the thermostat. When under this condition the vent blower should not run and the furnace should not light. Turn off the thermostat and electrical power and then reconnect the black wire to the vent blower motor.

COMPONENT PARTS

The following is a description of the furnace components. For their location refer to Figure 5.6.

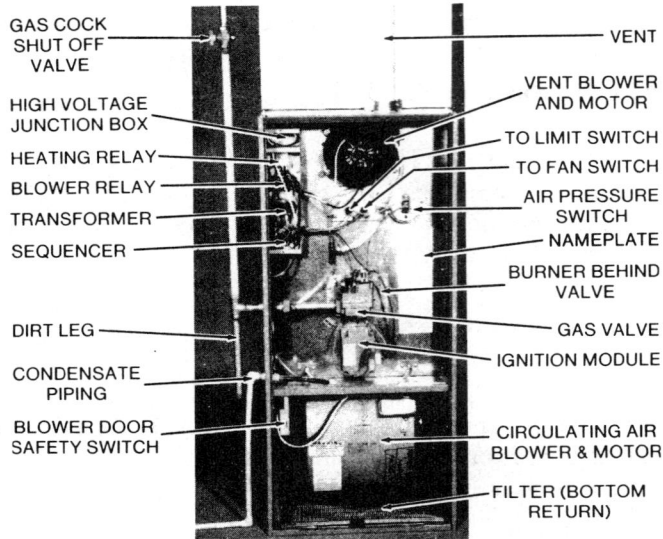

Figure 5.6 Furnace component location. (Courtesy of The Coleman Company.)

Gas Valve

The gas valve is a redundant-type valve. There are two operators that must open to allow gas to flow to the burner. Since the furnace cannot be match lighted, there is no pilot position on the control knob. They operate on 24 V ac.

> **Caution:** Never short the terminals on the gas valve. To do so may damage the valve or burn out the heat anticipator on the thermostat (see Figure 5.7).

Vent Motor

The vent motor is a 115-V ac motor that supplies combustion air to the burner. Air for combustion is pulled in through the louvers on the front panel, over

Figure 5.7 Gas valve. (Courtesy of The Coleman Company.)

the burner through the heat exchanger, and discharged out through the vent pipe (see Figure 5.8).

Figure 5.8 Vent motor. (Courtesy of The Coleman Company.)

Pressure Switch

The pressure switch is an air-actuated switch with normally open contacts. These contacts are connected to the tube that runs from the switch up to the vent blower housing. When the vent motor starts, the negative pressure on the tube closes the contacts on the pressure switch. If the flue should become obstructed, the contacts on this switch will not operate (see Figure 5.9).

A low-pressure reading indicates a severely restricted flue or combustion air inlet. A properly operating vent motor assembly should produce well over 0.5 in. wc negative pressure.

Figure 5.9 Pressure switch. (Courtesy of The Coleman Company.)

Ignition Module

The ignition module contains the timing circuits that produce the high-voltage spark to ignite the burner. At the same instant that the spark starts, the gas valve and the blower sequencer are energized. After the third try for ignition, if the burner is not ignited, the module locks out the system and it will stay locked out until the system is reset (see Figure 5.10).

Figure 5.10 Ignition module. (Courtesy of The Coleman Company.)

Heating Relay

The heating relay has three sets of contacts: two normally open sets and one normally closed set (see Figure 5.11). One set of normally open contacts controls the electrical power to the vent motor. The other set of normally open contacts controls the electrical power through the normally closed limit switch and normally open air pressure switch to the ignition module. The set of normally closed contacts provides the power to the blower relay for blower operation in either the cooling or continuous mode.

Figure 5.11 Heating relay. (Courtesy of The Coleman Company.)

Blower Relay

This is a double-pole, double-throw relay. One set of normally open contacts is wired to the cooling speed of the blower motor. The normally closed contacts are wired to the heating speed of the blower motor (see Figure 5.12).

Figure 5.12 Blower relay. (Courtesy of The Coleman Company.)

Blower Sequencer

The sequencer is energized by the ignition module at the same instant that the gas valve is energized. Approximately 30 seconds after the sequencer is energized the normally open contacts close starting the blower motor. These contacts bypass the thermal fan switch. The purpose of this switch is to start the blower with a minimum of delay after the burner is ignited (see Figure 5.13).

Blower Motor

The blower motor is a 115-V ac, multispeed permanent split-capacitor type. It is factory wired for high speed in heating operation. The speed for cooling is

Figure 5.13 Blower sequencer. (Courtesy of The Coleman Company.)

optional and should be selected to match the cooling load. The heating speed can be adjusted to either of the two highest blower motor speeds.

Transformer

The purpose of the transformer is to reduce the 115-V ac primary voltage to 24 V ac, which the service technician can work with easily and safely. The transformer is rated at 40 VA output with sufficient capacity for add on air conditioning.

Fan Switch

This is a thermally actuated normally open switch that closes when the heat exchanger temperature reaches the switch operating temperature. It opens after the burner is out and the heat exchanger cools off (see Figure 5.14).

Figure 5.14 Fan switch. (Courtesy of The Coleman Company.)

Limit Control

The limit control is a thermally actuated normally closed switch (see Figure 5.15). It opens to deenergize the ignition module and close the gas valve when the heat exchanger temperature becomes excessive.

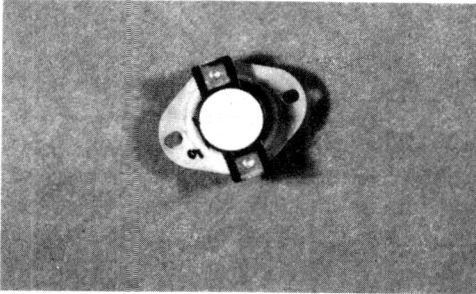

Figure 5.15 Limit control. (Courtesy of The Coleman Company.)

Blower Door Safety Switch

This switch is located inside the blower compartment and controls the electrical supply to the furnace electrical circuits. When the blower door is put in place, the switch is activated, allowing the furnace to operate. The switch is designed to prevent furnace operation if the blower door is removed and inadvertently not reinstalled, thus preventing the possibility of the blower creating a negative pressure in the furnace enclosure.

Burner

The burner is a monoport type with a stainless steel flame spreader and an adjustable primary air (see Figure 5.16).

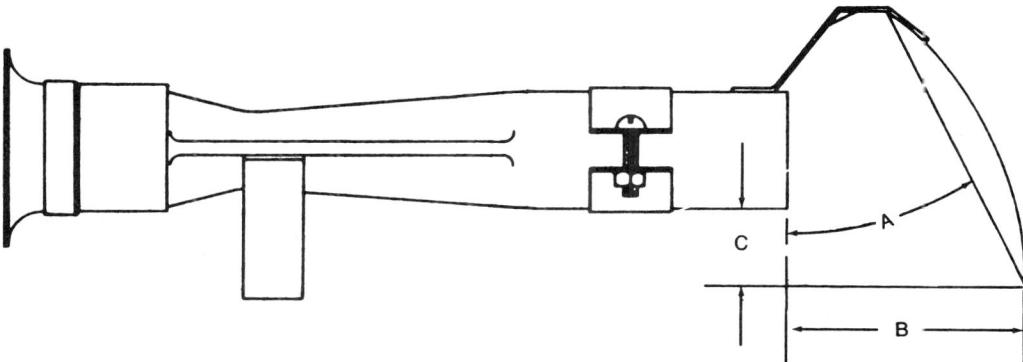

Figure 5.16 Burner. (Courtesy of The Coleman Company.)

VENT SYSTEM

The furnace described in these instructions is an induced-draft type and is vented by an electrically powered vent blower. The vent piping operates under a slight positive pressure; therefore, it is important that the vent piping be installed properly to ensure that there will be no leakage of the products of combustion. Refer to Chapter 1 of this manual for details of the vent installation. It is essential that the furnace be vented properly and that the special requirements listed below be followed.

> **Caution:** Failure to follow the following requirements and recommendations can cause unsatisfactory furnace operation, fire, explosion, or asphyxiation.

Special Requirements

The following is a list of the special requirements for this furnace:

1. The furnace must be vented using only schedule 40 PVC pipe. All joints should be sealed securely and any horizontal pipe should slope slightly toward the furnace. This is to ensure that any condensation in the pipe will drain to the furnace and that flue gases will travel upward and out.
2. The furnace must be vented separately from any other appliance, with the vent running continuously from the furnace to the outside atmosphere. Do not vent this furnace into a common flue or vent with any other appliance, such as a water heater, space heater, wood-burning stove or fireplace, or clothes dryer.
3. Do not vent this furnace into a masonry chimney or any other all-fuel-type chimney.

GAS CONVERSION

Coleman furnaces can be converted to LP gas by using a conversion kit that is available from The Coleman Company, Inc. The conversion kit consists essentially of instructions, burner orifice, two stickers, and a spring for converting a White Rodgers valve. See the installation instructions packed with each kit for complete instructions.

Service Procedures

Before removing components for inspection or service, carefully observe how these components are installed, electrical connections made, and attachment means employed.

All components and assemblies must be reinstalled in the same manner and position as they were originally.

Damaged or defective components must be replaced. Only original equipment components, or substitute components authorized by The Coleman Company may be used in the repair of these furnaces.

Vent Motor Assembly

The vent motor assembly consists of a motor, impeller, and housing. If any part of the assembly fails or is defective, the entire assembly must be replaced. If the vent motor does not run, check for 115 V between No. 1 on the heating relay and the white wire to the motor. If voltage is present and the motor does not run the assembly must be replaced. If voltage is not present, use the troubleshooting charts in the back of this chapter to determine the cause.

To remove the vent motor assembly.   Use the following steps when removing the vent motor assembly:

1. Turn off the electrical power to the furnace and unplug the power wires from the motor.
2. Disconnect the 2-in. CPVC pipe from the outlet of the vent motor assembly.
3. Remove the mounting screws that secure the vent motor assembly to the furnace.
4. Pull the vent motor assembly straight out. The vent motor assembly is installed and sealed over a protuding flange on the furnace. A slight forward pressure may be necessary to remove the assembly.

To install the vent motor assembly.   Use the following steps to install the vent motor assembly:

1. Remove any sealant that was left on the furnace flange when the vent motor assembly was removed.
2. Run a small bead of RTV silicone cement in the groove of the vent motor assembly that installs over the furnace flange outlet.
3. Line up the assembly with the mounting holes and press inward slightly.
4. Reinstall the mounting screws and connect the electrical wires to the motor.
5. Reinstall the 2-in. CPVC vent pipe to the outlet of the vent motor assembly.
6. Turn on the furnace and operate it for 10 minutes to be sure that there are no condensate leaks.

Burner Assembly

The burner should be inspected for any accumulation of dust or lint. Use the pressure side of a vacuum cleaner to blow out any dust or lint. The burner must be removed to inspect the ignition electrode assembly and the heat exchanger.

To remove the burner assembly. Use the following steps to remove the burner assembly:

1. Turn off the gas supply to the furnace.
2. Disconnect the gas line to the gas valve.
3. Disconnect the low-voltage wires to the gas valve.
4. Remove the four screws holding the gas valve mount to the furnace vestibule.
5. Remove the four screws holding the burner assembly to the furnace vestibule.

> **Caution:** Remove the burner slowly, being careful not to damage the electrode assembly.

Ignition Electrode Assembly

Check the condition of the electrode rods: They should show no sign of serious deterioration, scaling, or carbon build up. Check the dimensional relationship between the rods and the burner (see Figure 5.17). Minor adjustments can be made by carefully bending the rods slightly. Take care not to damage the porcelain insulators.

Combustion Chamber

Using an inspection mirror and a flashlight, inspect the inside of the heat exchanger for any scaling, soot deposits, or metal fatigue. If any soot deposits are found, the heat exchanger must be cleaned. Soot is caused by improper burner adjustment, which must be corrected when the burner is reinstalled. If any holes are found, the heat exchanger must be replaced.

Following the inspection and cleaning reinstall all component parts in the reverse order of their removal.

IGNITOR ELECTRODE
FLUSH ± 1/16 WITH
SIDE OF BURNER

1/2" ± 1/16

1/8" GAP BETWEEN
IGNITOR ELECTRODE AND
GROUND ELECTRODE TIPS

90°

AIR SHUTTER

AIR SHUTTER ADJUSTMENT
ROD

AIR SHUTTER
LOCKING
SCREW

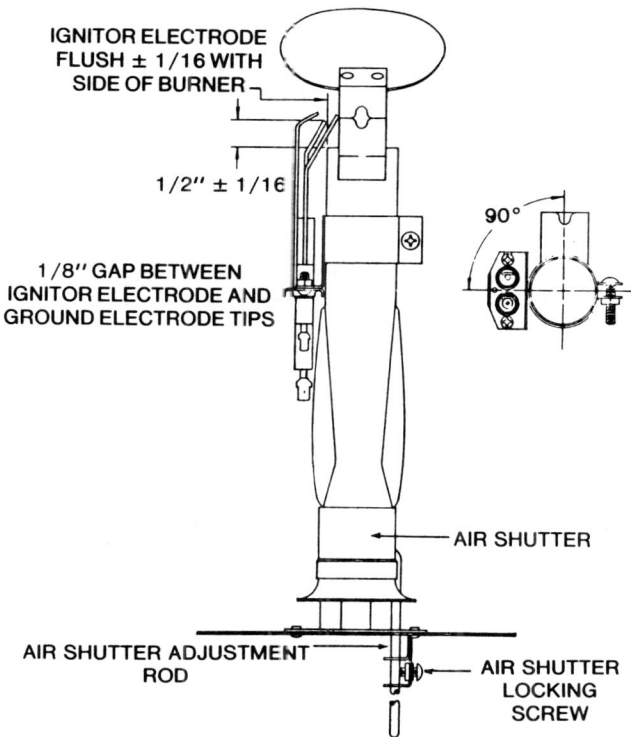

Figure 5.17 Ignitor electrode and
burner dimensions. (Courtesy of The
Coleman Company.)

Pressure Switch

The pressure switch is a normally open switch that is actuated by the negative
pressure developed by the vent blower. It is factory set to close at −0.35 in. wc
negative pressure and to open at −0.30 in. wc negative pressure. If there is any
obstruction in the flue outlet or combustion air inlet the switch will not close.

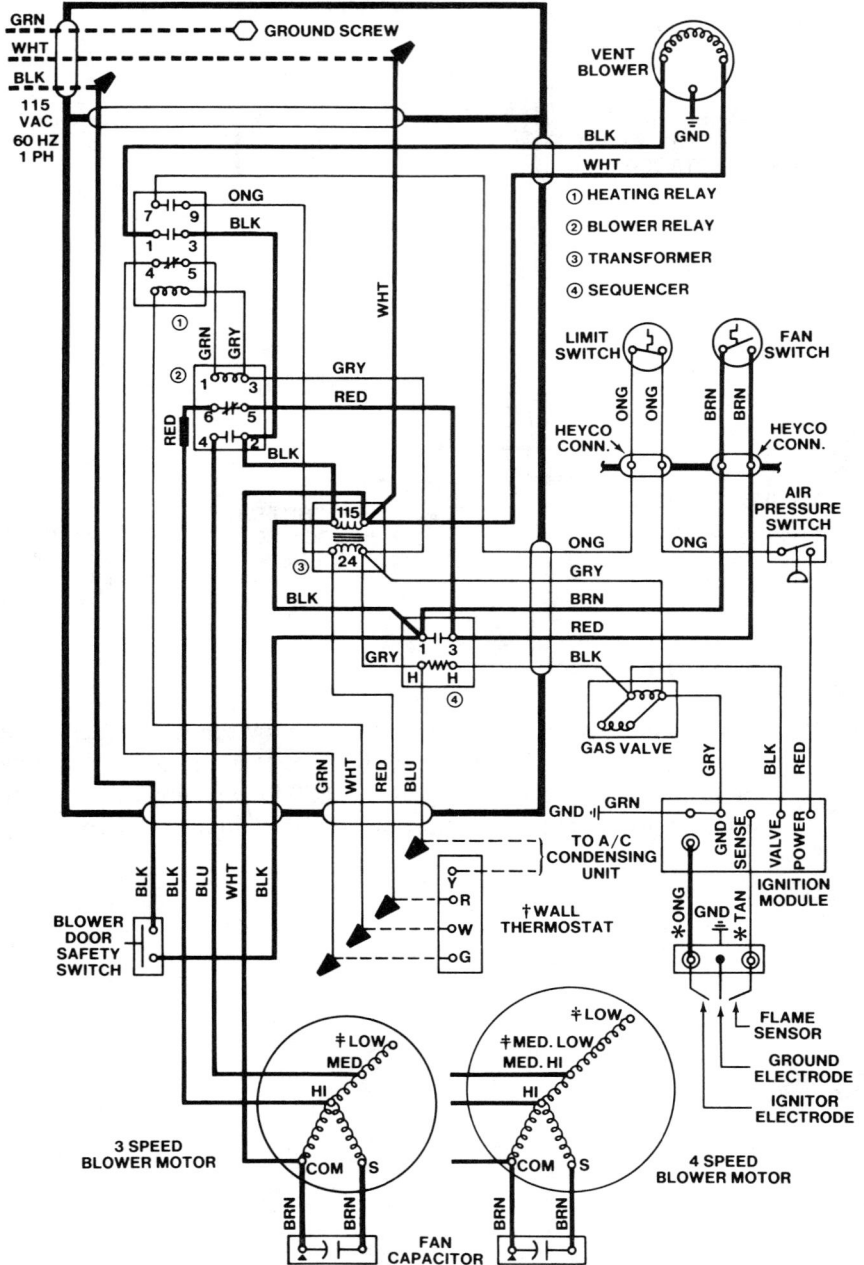

THE COLEMAN T.H.E. 90 GAS FURNACE (''———'' AND ''A'' MODEL) UNIT WIRING DIAGRAMS

90 SERIES
COLEMAN GAS FORCED AIR FURNACE
USE ONLY 115 VAC 60 HZ 1 PH
LESS THAN 12 AMPS   MAX. OVERCURRENT PROTECTION 15 AMPS

GRN
WHT
BLK
115 VAC 60 HZ 1 PH

GROUND SCREW

VENT BLOWER

BLK   GND
WHT

ONG
BLK

① HEATING RELAY
② BLOWER RELAY
③ TRANSFORMER
④ SEQUENCER

7 9
1 3
4 5

GRN
GRY
RED

GRY
RED

BLK

WHT

LIMIT SWITCH

FAN SWITCH

ONG ONG   BRN BRN

HEYCO CONN.   HEYCO CONN.

AIR PRESSURE SWITCH

115
24

BLK
GRY
BRN
RED
BLK

ONG   ONG

GRY

GRY 1 3 BLK
H   H

GAS VALVE

GRN WHT RED BLU

GND GRN

TO A/C CONDENSING UNIT

Y
R
W
G

†WALL THERMOSTAT

BLK BLK BLU WHT BLK

BLOWER DOOR SAFETY SWITCH

GRY GND SENSE VALVE POWER

IGNITION MODULE

*ONG GND *TAN

FLAME SENSOR
GROUND ELECTRODE
IGNITOR ELECTRODE

‡LOW
MED.
HI
COM S

3 SPEED BLOWER MOTOR

‡LOW
‡MED. LOW
MED. HI
HI
COM S

4 SPEED BLOWER MOTOR

BRN BRN   BRN BRN

FAN CAPACITOR

FACTORY INTERNAL WIRING SHOWN SOLID

SERVICE CHARTS

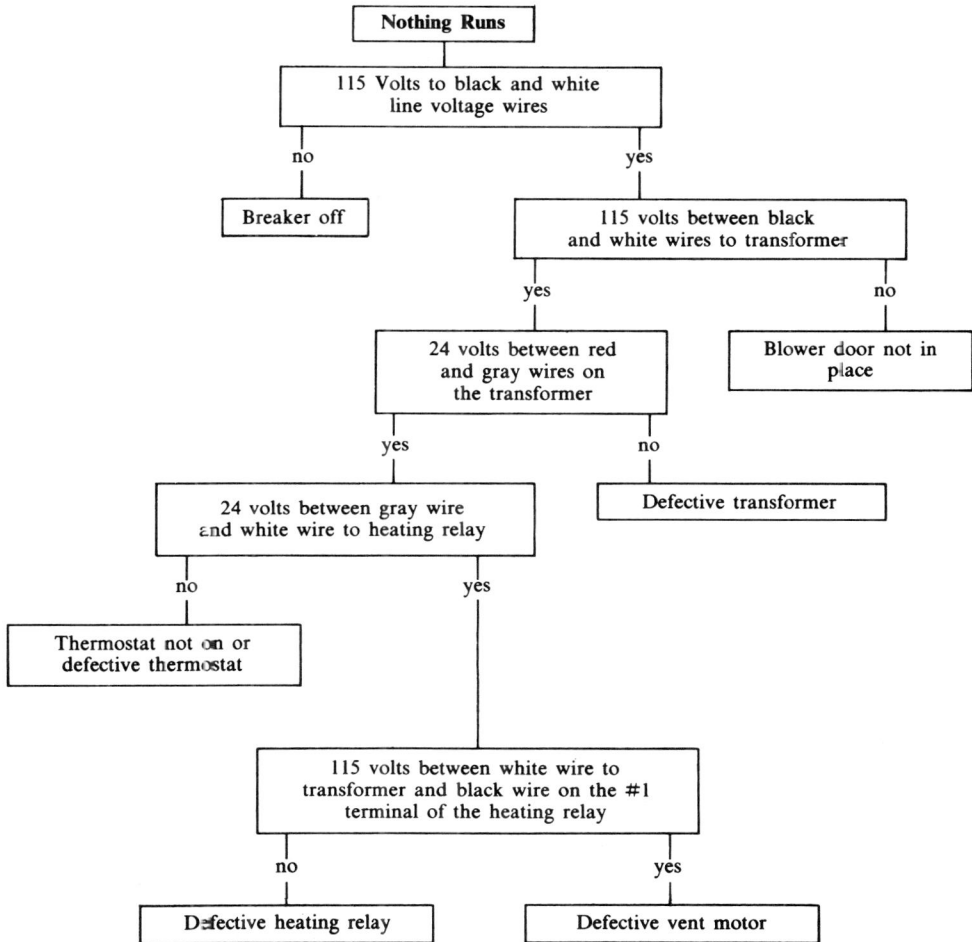

```
                        ┌──────────────────┐
                        │   Nothing Runs   │
                        └──────────────────┘
                        ┌──────────────────────────┐
                        │ 115 Volts to black and white │
                        │      line voltage wires      │
                        └──────────────────────────┘
                  no                              yes
          ┌──────────────┐              ┌────────────────────────┐
          │  Breaker off │              │  115 volts between black │
          └──────────────┘              │ and white wires to transformer│
                                        └────────────────────────┘
                                   yes                         no
                        ┌────────────────────┐      ┌────────────────────┐
                        │ 24 volts between red │      │ Blower door not in │
                        │  and gray wires on   │      │       place        │
                        │   the transformer    │      └────────────────────┘
                        └────────────────────┘
                   yes                      no
          ┌────────────────────────┐   ┌────────────────────┐
          │ 24 volts between gray wire │   │ Defective transformer │
          │ and white wire to heating relay│   └────────────────────┘
          └────────────────────────┘
          no                     yes
  ┌────────────────────┐
  │ Thermostat not on or │
  │ defective thermostat │
  └────────────────────┘
                        ┌──────────────────────────────┐
                        │ 115 volts between white wire to │
                        │ transformer and black wire on the #1 │
                        │   terminal of the heating relay   │
                        └──────────────────────────────┘
                   no                          yes
          ┌────────────────────────┐   ┌────────────────────┐
          │ Defective heating relay │   │ Defective vent motor │
          └────────────────────────┘   └────────────────────┘
```

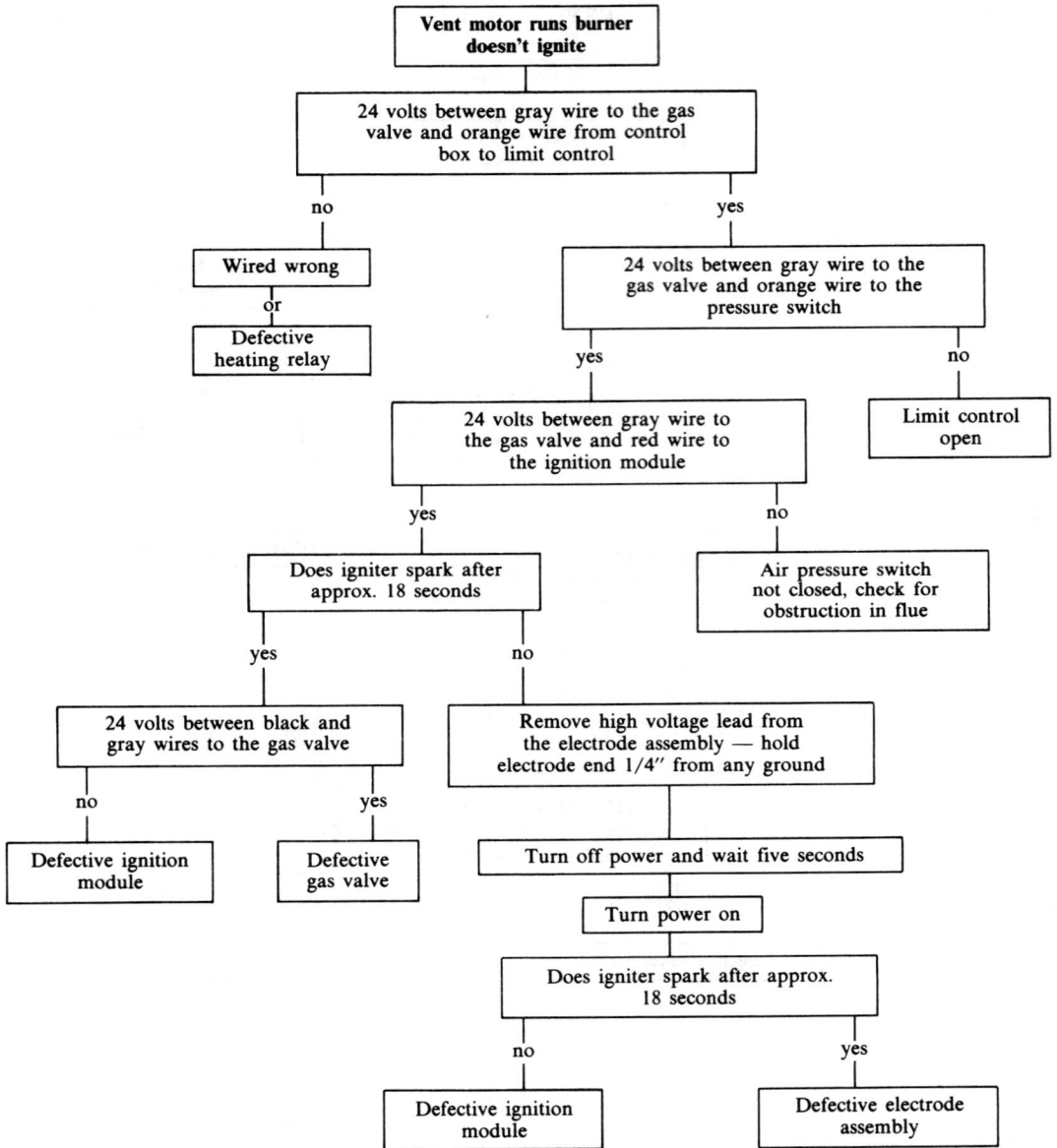

```
                        ┌─────────────────────────┐
                        │  Vent motor runs burner  │
                        │      doesn't ignite      │
                        └─────────────────────────┘
                                    │
                        ┌─────────────────────────────┐
                        │ 24 volts between gray wire to the gas │
                        │ valve and orange wire from control    │
                        │      box to limit control             │
                        └─────────────────────────────┘
                    no                              yes
```

Vent motor runs burner doesn't ignite

24 volts between gray wire to the gas valve and orange wire from control box to limit control

- no → **Wired wrong** or **Defective heating relay**
- yes → **24 volts between gray wire to the gas valve and orange wire to the pressure switch**
  - no → **Limit control open**
  - yes → **24 volts between gray wire to the gas valve and red wire to the ignition module**
    - no → **Air pressure switch not closed, check for obstruction in flue**
    - yes → **Does igniter spark after approx. 18 seconds**
      - yes → **24 volts between black and gray wires to the gas valve**
        - no → **Defective ignition module**
        - yes → **Defective gas valve**
      - no → **Remove high voltage lead from the electrode assembly — hold electrode end 1/4″ from any ground**
        - **Turn off power and wait five seconds**
        - **Turn power on**
        - **Does igniter spark after approx. 18 seconds**
          - no → **Defective ignition module**
          - yes → **Defective electrode assembly**

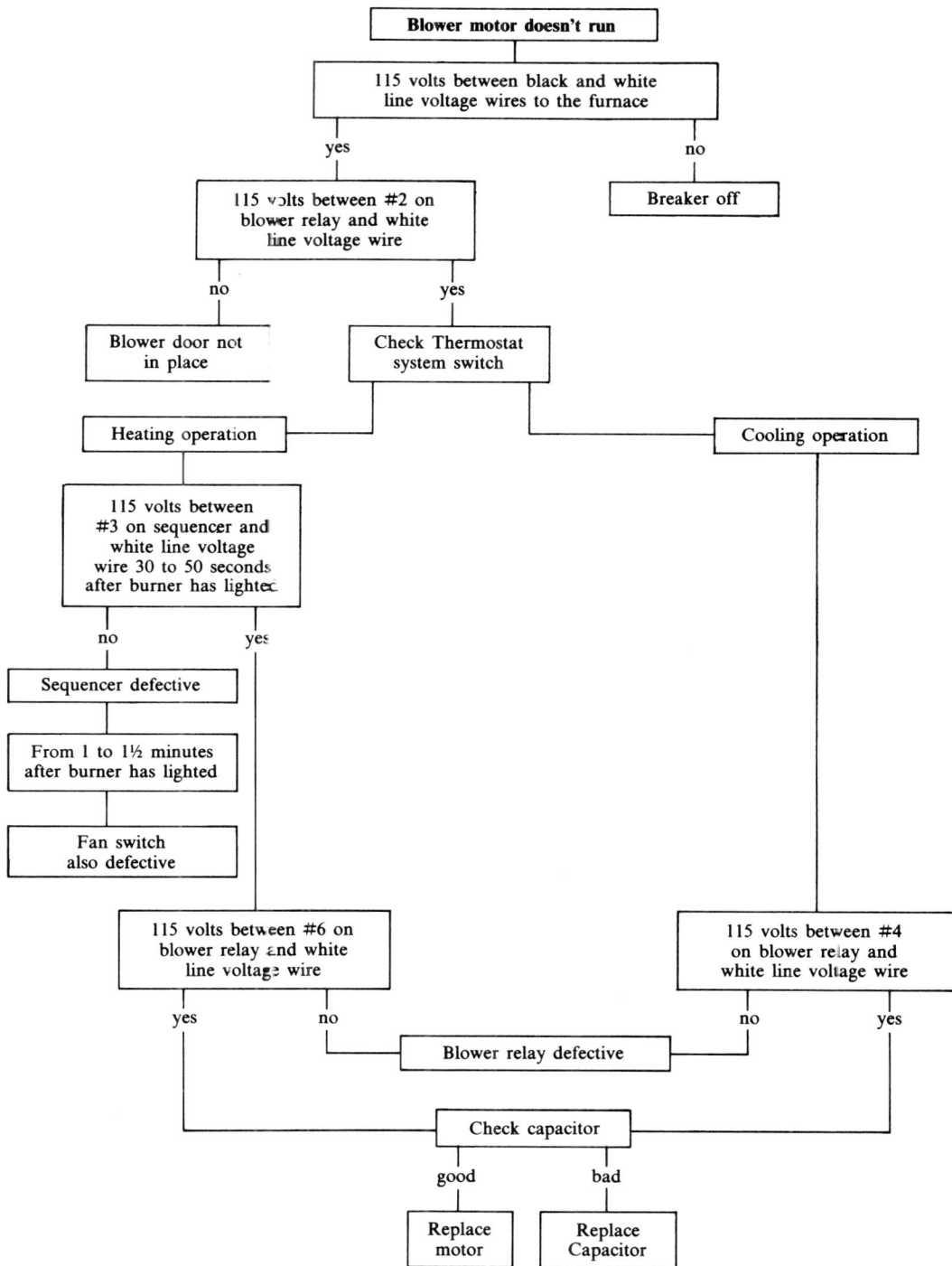

Blower motor doesn't run

115 volts between black and white line voltage wires to the furnace

— yes →
- 115 volts between #2 on blower relay and white line voltage wire
- — no → Blower door not in place
- — yes → Check Thermostat system switch

— no →
- Breaker off

Check Thermostat system switch:
- Heating operation
- Cooling operation

Heating operation:
- 115 volts between #3 on sequencer and white line voltage wire 30 to 50 seconds after burner has lighted
  - no → Sequencer defective → From 1 to 1½ minutes after burner has lighted → Fan switch also defective
  - yes → 115 volts between #6 on blower relay and white line voltage wire
    - yes → Check capacitor
    - no → Blower relay defective

Cooling operation:
- 115 volts between #4 on blower relay and white line voltage wire
  - no → Blower relay defective
  - yes → Check capacitor

Check capacitor:
- good → Replace motor
- bad → Replace Capacitor

191

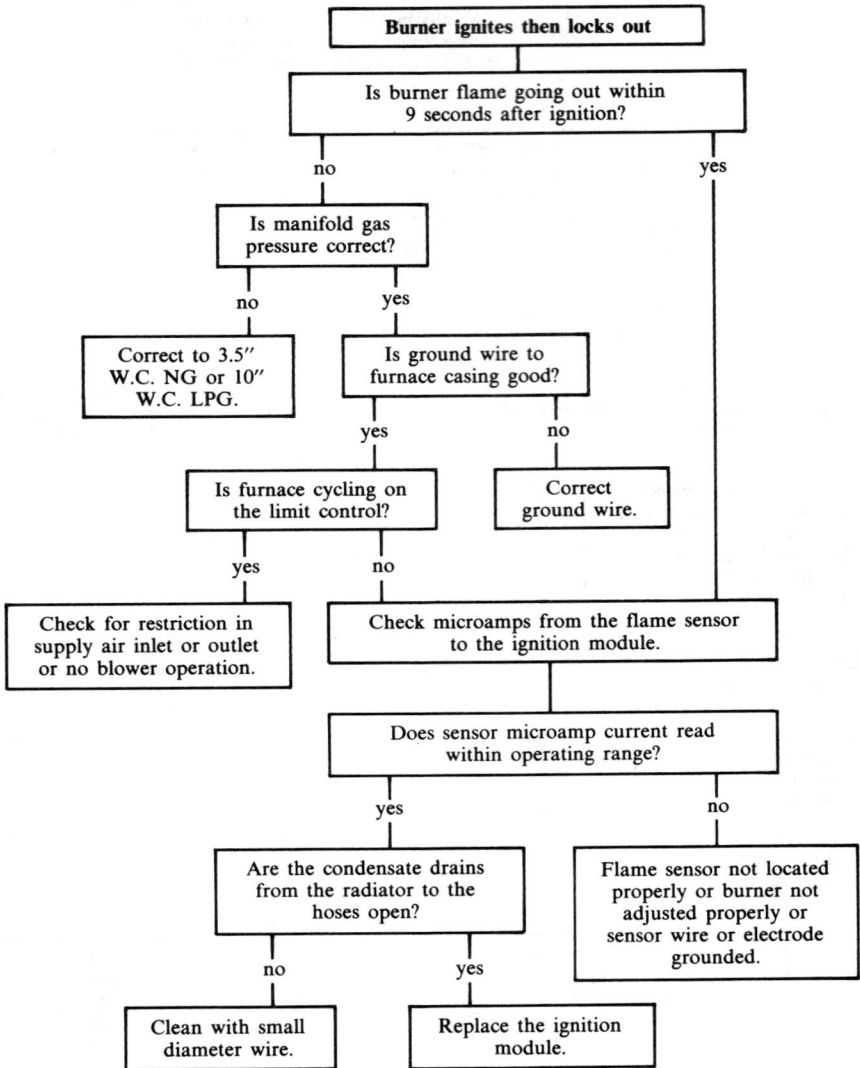

6

The Coleman T.H.E. Model B High-Efficiency Gas Furnace

These instructions are for the use of qualified individuals specifically trained and experienced in the installation of this type of equipment and related system components.

Installation and service personnel are required by some states to be licensed. Persons not qualified shall not attempt to install this equipment or interpret these instructions.

> **Warning:** Improper installation may damage the equipment or create a hazard that could result in a fire, explosion, or asphyxiation, and will void the warranty.

Note: The words *shall* and *must* indicate a requirement that is essential to safety and safe product performance.

The words *should* and *may* indicate a recommendation or advice that is not essential and not required but that may be useful or helpful.

APPLICATION

The following is a description of the application requirements for this furnace.

Furnace Certification and Usage

The furnace models described in these instructions are design certified by the American Gas Association to be in compliance with the American National Standard Z21.47b–1986.

These furnaces are the forced-air type and may be utilized for indoor installation in buildings constructed on the site, or manufactured buildings (modular only). These furnaces are not certified for installation in mobile homes.

Municipal, State, and Federal Codes

The installer must conform to all state and local building codes when installing these appliances. In the absence of state and local codes, these furnaces must be installed in accordance with the latest issue of the following:

National Electrical Code, ANSI/NFPA 70-1984.
National Fuel Gas Code, ANSI Z223.1–1984.

> **Warning:** During installation, cover the flue opening to prevent construction debris from entering the vent blower. Failure to cover the flue opening could cause furnace malfunction, resulting in fire or explosion.

Furnace Sizing and Duct System Design

Consideration should be given to the heating capacity required and also the air quantity (CFM) required if an air conditioning system is to be installed along with the furnace or at some future time. These factors can be determined by calculating the heat loss and the heat gain of the home or structure.

If these calculations are not performed and the high-efficiency furnace is oversized, the following may result:

1. Short cycling of the furnace.
2. Wide temperature fluctuations from the thermostat setting.
3. Reduced overall operating efficiency of the furnace.

The supply and return duct systems must be of adequate size and designed such that the furnace will operate within the designed air temperature rise range and not exceed the maximum designed static pressure. These values are listed in the specification table of these instructions and on the unit rating plate (see Table 6.1).

TABLE 6.1 SPECIFICATIONS (COURTESY OF THE COLEMAN COMPANY)

Model	Nat. L.P.	2940B666 2940B669	2960B666 2960B669	2970B766 2970B769	2985B766 2985B769	2960B766 2960B769
Unit rating	Input: 0-2,000 ft. elevation	45,000	65,000	80,000	95,000	65,000
Btu/hr.	High altitude	For elevations above 2,000 ft. reduce input 4% for each 1,000 ft. of elev. above sea level				
Air temperature rise range °F		20-50	25-55	30-60	35-65	25-55
Designed max. outlet air temp. °F		145	150	155	160	150
Max. external static pressure in. W.C.		.5†				
Vent pipe		2 in. diameter * schedule 40 PVC or CPVC††				
Vent pipe fittings		2 in. diameter schedule 40 or 80 PVC or CPVC††				
Condensate pipe and fittings		1/2 in. CPVC or PVC				
Gas connection		1/2 in. FPT				
Electrical service		115 VAC 60 Hz 1 Ph				
Minimum distance to combustible materials—inches Front Back Plenum-top Floor Sides Vent		6 0 1 Combustible 0 0				
Filter (furnished)		16 × 25 × 1/2		20 × 25 × 1/2		
Block-off panel kit (optional)		2940-1451		2632-1451		

*3 in. diameter must be used with model 2985; except, furnace connection and vent termination are 2 in.
†Includes the combined total of the return duct, the supply duct, plus the pressure drop across a coil when installed.
††First two feet must be CPVC.

The information, values, and other data necessary for heat loss, heat gain, and duct system design may be found in the *ASHRAE Handbook of Fundamentals,* 1972 edition, or in other publications that are recognized by municipal, state, and federal code authorities.

When used with air conditioning.   The following precautions must be observed when this furnace is to be used in combination with an air conditioning system:

1. When a single (common) duct is used:
   a. A plenum-type cooling coil must be installed on the air discharge side downstream from the furnace, or
   b. A coil-blower-type cooling coil must be installed in parallel with and isolated from the furnace, or
   c. A self-contained air conditioning unit must be in parallel with and isolated from the furnace.

> **Warning:** Dampers must be installed when a coil-blower or a self-contained unit is used to prevent conditioned air from coming in contact with the heat exchanger to avoid moisture condensation and rust, which can allow the products of combustion to be circulated into the living area by the furnace blower, resulting in possible asphyxiation.

   If the dampers are manually operated a means must be provided to prevent either the furnace or the air conditioning unit from operating unless the dampers are in the full heat or cool position.

2. If two duct systems are used:
   If two complete or partial duct systems are used, as could be the case with a coil-blower or a self-contained air conditioning unit, the furnace and the air conditioning unit should be controlled by a single combination heat and cool thermostat, which will prevent the furnace and the air conditioning unit from operating simultaneously.

> **Caution:** If separate heating and cooling thermostats are used, a manually operated electrical interlock switch must be installed to prevent simultaneous operation of both systems and avoid a possible hazardous condition due to overheating of the conditioned space.

Ignition System Energy Consumption

The electrical energy consumed by the electric ignition system, not including the gas valve, during continuous operation is 2.5 W or less.

The efficiency rating of the furnace is a product thermal efficiency rating determined under continuous operating conditions independent of any installed system.

INSTALLATION PROCEDURE

The following steps must be observed when installing this type of furnace:

Notice to the Installer Before the initial operation of this furnace the installer must complete the following steps:
1. Remove the blower housing from the furnace.
2. Remove the shipping strap from the left end of the blower housing.
3. Remove the cardboard shipping pad from the bottom of the furnace blower compartment.
4. Reinstall the blower housing in the furnace.
5. After completing steps 1 through 4, remove the NOTICE TO INSTALLER sticker from the front of the furnace.

Locations and Clearances

The minimum clearances between the furnace, vent system, etc., and combustible materials are listed in the specification tables of these instructions and on the unit rating plate (see Table 6.1).

Installations on combustible flooring. This furnace shall not be installed directly on carpeting, tile, or other combustible material, other than wood flooring.

Clearance for lighting and service. Adequate clearance must be provided for lighting and maintenance. A minimum of 24 in. clearance should be provided between the front of the furnace and any opposite wall. If the furnace is in line with a door, a minimum of 6 in. clearance must be provided from the front of the furnace to the door.

If the furnace is to be installed in a close clearance closet, the door should be of adequate size to allow for removal of the furnace should it become necessary.

Installation in residential garages. When the furnace is to be installed in a residential garage it must be located such that it will be protected against vehicular damage. The furnace must be installed such that the burner and electric ignition control module are a minimum of 18 in. above the floor.

Installation in an unconditioned space. This is a condensing gas furnace; therefore, if the temperature in an unconditioned space ever reaches freezing,

32 °F, or below, the condensate will freeze in the condensate tubes. If this happens, the condensate will fill the condensing radiator until it blocks the flow of flue gases which in turn will shut down the furnace. The furnace will not operate until the condensate thaws and unblocks the flow of flue gases through the condensing radiator. Do not install this furnace in any unconditioned space that may have temperatures that reach 32 °F or below.

Ventilation and Combustion Air

Provide ventilation and combustion air in accordance with Section 5.3, Air for Combustion and Ventilation, ANSI Z223.1–1984, or applicable provisions of the local building codes.

> **Warning:** Adequate ventilation and combustion air must be provided to ensure satisfactory and safe operation of the furnace. Air openings in the front casing, front panel, and vestibule on the top panel must not be obstructed. Failure to observe this recommendation could result in asphyxiation.

Do not store halogen-emitting substances, such as laundry bleach and detergent, cleaning fluids, spray can propellants, and solvents, in the vicinity of this appliance. The air used by the burner for combustion must be free of halogens to avoid possible corrosion to the heating surfaces, which could result in asphyxiation.

Installations in a confined space.   If the unit is to be installed in a confined space such as a small closet or room, provisions must be made for supplying combustion and ventilation air to the space surrounding the furnace (see Figure 6.1). Two openings of equal area must be provided: one commencing within 12 in. of the ceiling and one commencing within 12 in. of the floor of the confined space. The upper opening shall always be above the top of the furnace casing. The lower opening in the floor, side wall, or door shall be located below the level of the burner in the furnace.

All Air from Inside the Building.   The total free area of each opening must be at least 1 in.² for each 1000 Btuh of furnace input, but not less than 100 in.².

All Air from Outdoors.   When communicating directly with the outdoors or through vertical ducts, the total free area of each opening must be at least 1 in.² for each 4000 Btuh of furnace input.

When communicating with outdoors through horizontal ducts, the total free area of each opening must be at least 1 in.² for each 2000 Btuh of furnace input.

OPENING FOR
VENTILATION AIR ———————— OR ———

OPENING FOR
COMBUSTION AIR ———————— OR ——

AIR OPENINGS - CONFINED SPACE

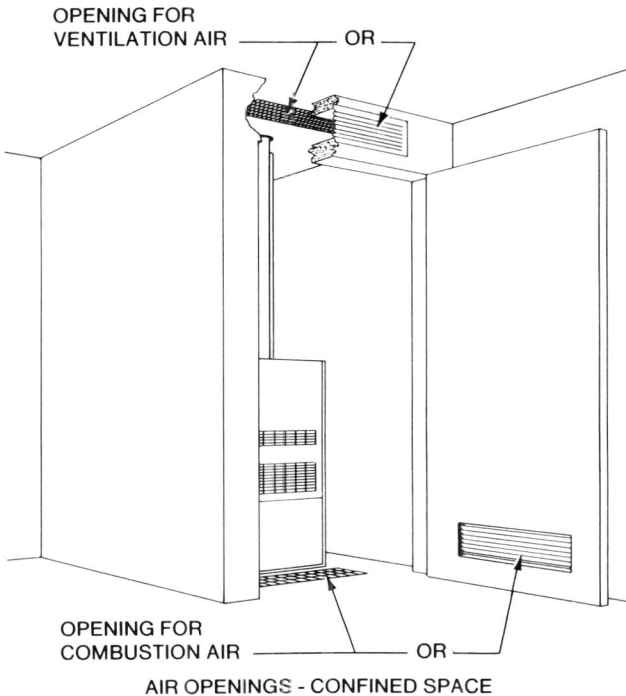

Figure 6.1 Combustion air opening locations. (Courtesy of The Coleman Company.)

When ducts are used, they must be of the same cross-sectional area as the free area of the openings to which they connect. The minimum dimension of rectangular air ducts must not be less than 3 in.

Installations in an unconfined space.   In unconfined spaces in a building, infiltration normally is adequate to provide air for combustion and ventilation.

Installation of Plenum
or Air Conditioning Coil Cabinet

Because it may be necessary to remove the plate covering the top of the furnace vestibule compartment for a periodic maintenance inspection or service, observe the following procedure. Remove the plate, if the plenum has a horizontal flange at its bottom, and reinstall the plate over the top of the plenum flange. Drill the necessary holes and attach the plate to the plenum flange and furnace (see Figure 6.2).

If an air conditioning coil is being installed, remove and discard the back screws securing the plate to the vestibule. Install the coil cabinet over the top of the plate.

REMOVE 2
SCREWS FOR
A/C CABINET
APPLICATION

PLATE

Figure 6.2 Plenum or A/C coil cabinet installation. (Courtesy of The Coleman Company.)

Do not exceed 0.5 in. wc total static pressure. See duct system design presented earlier in this chapter.

Return Air and Filters

The return air may be brought in the bottom or on either side of the furnace. Lances on the casing side locate the return air openings.

When a bottom return air inlet is used, the filter is installed in the blower compartment over the air inlet opening and secured in place with the remaining clip provided. When a side return air inlet is used, the filters cannot be mounted inside the casing. The filters must be mounted in an external frame on the side of the furnace or in the return air duct and be accessible for cleaning or replacement. One or more return grill-filter frames may be used when side inlet returns are used.

> **Important Notice:** A solid block-off metal panel must be in place to block the bottom opening in the furnace when side or rear return air ducts are used. Failure to block this opening could cause products of combustion to be circulated into the living space and create a potentially hazardous condition. See the specifications table (Table 6.1) for an optional block-off panel kit.

> **Caution:** When the furnace is installed in a closet or other confined space, and a side inlet duct is used, the duct must be sealed to the furnace and extended to the conditioned space top to prevent any communication between the space or room in which the furnace is installed, which could result in asphyxiation. Do not exceed 0.5 in. wc total static pressure. See duct system design presented earlier in this chapter.

Gas Piping and Supply Pressures

Before installing the gas piping, check with the local code authorities for specific requirements concerning gas piping.

In the absence of local codes follow the recommendations in the National Fuel Gas Code, ANSI Z223.1–1984, for gas piping materials, pipe sizing, and the requirement for installation. It is recommended that a gas-cock shutoff valve be installed in the gas supply line outside the furnace casing, where it is readily accessible, as close to the furnace as possible.

A 1/8-in. NPT (plugged) pressure tap, for a test gauge connection, must be installed in the gas supply line immediately upstream of the furnace.

Install a dirt leg at the bottom of any vertical riser, or drop, as close to the furnace as possible, to collect moisture and foreign material. Install a ground joint union just ahead of the gas control valve (see Figure 6.3).

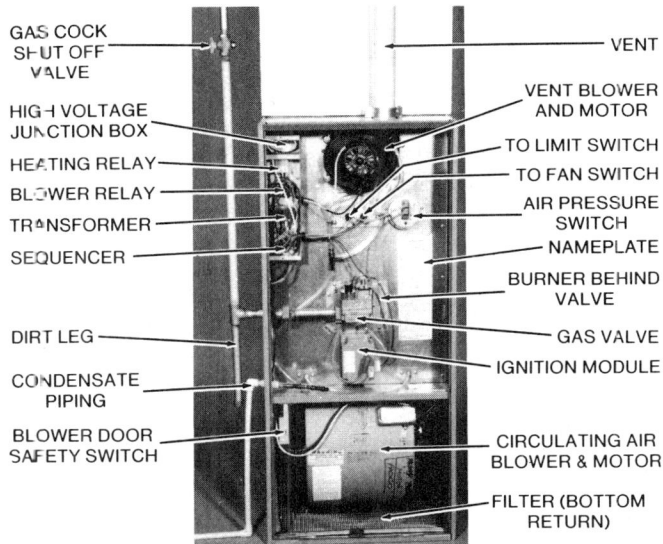

Figure 6.3 Furnace component arrangement. (Courtesy of The Coleman Company.)

When making the connection at the gas control valve, use a wrench on the inlet side of the valve to prevent any possible twisting of the valve body, which could cause damage and leaks. The connection sizes are shown in Table 6.1. When making up pipe joints, use pipe joint thread compound that is resistant to natural and LP gases.

Caution: During the pressure testing of the gas supply piping system, observe the following procedure to avoid damage to the appliance, fire, explosion, or asphyxiation.

    a. If the test pressure is equal to or less than one-half in. psig, isolate the furnace by closing its individual manual shutoff valve.

    b. If the test pressure is greater than one-half in. psig, the furnace and its individual shutoff valve must be disconnected from the gas supply piping system.

Following the installation of the piping, first ensure that the gas control knob on the gas valve is in the off position, then pressurize the piping system with gas. Next, purge the piping system of air. Be sure that any cigarettes, open flames, or other ignition sources are extinguished and that there is adequate ventilation in any confined space.

Thoroughly check the piping system for any leaks.

Caution: Never use an open flame: Fire or explosion could occur. Since some leak solutions including soap and water may cause corrosion or stress cracking, the piping shall be rinsed with water after testing unless it has been determined that the leak test solution is noncorrosive.

The maximum and minimum gas supply pressure required at the inlet of the gas control valve is shown on the unit rating plate. When the furnace is in operation the inlet pressure must be within the limits shown.

Venting Combustion Gases

These furnaces are of the induced-draft type and are vented by an electrically powered vent blower. The vent piping operates under a slight positive pressure; therefore, it is important that the vent piping be properly installed to ensure that there will be no leakage of products of combustion. To ensure that there will be no leakage of the products of combustion, seal all vent pipe joints with a PVC/CPVC cement.

In some venting applications and certain geographical areas where mud daubers, birds, rodents, etc., can create blockage or restrictions in the vent pipe, it is advisable to install accesses in the PVC or CPVC pipe at strategic locations, allowing the pipe to be easily cleaned. The screw-in plug must fasten tight enough to prevent any condensate that might form on the pipe from leaking through the threads. Coleman recommends using tees with cleanout fittings and threaded plugs (see Figure 6.4).

HORIZONTAL VENT
ACCESS ABOVE
FURNACE

VERTICAL VENT
ACCESS ABOVE
FURNACE

CAUTION
INSTALL ACCESS TEE
IN A VERTICAL VENT
SO IT WILL NOT CATCH
AND HOLD CONDENSATE.

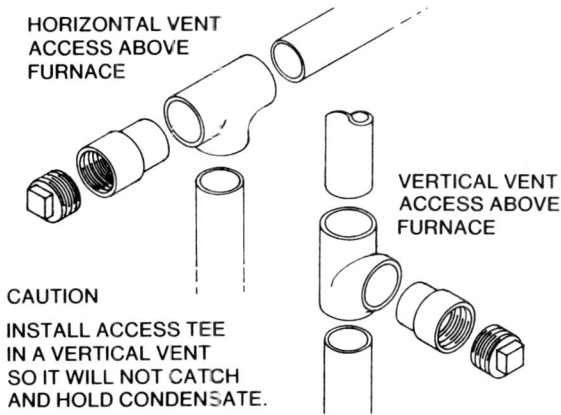

Figure 6.4 Vent pipe access arrangement. (Courtesy of The Coleman Company.)

Selecting the best vent arrangement. The vent system may be of two basic types:

1. A horizontal vent that terminates at the outer surface of an outside wall that is above grade. This type of venting must terminate with a 2-in.-diameter schedule 40 or 80 plastic PVC or CPVC tee for the vent terminal.
2. A vent arrangement that consists of a vertical riser or horizontal lateral connected to a vertical riser that passes through the interior of the structure and terminates above the roof with a 2-in. schedule 40 or 80 plastic PVC or CPVC return bend (two 90° elbows) or tee (see Figure 6.5).

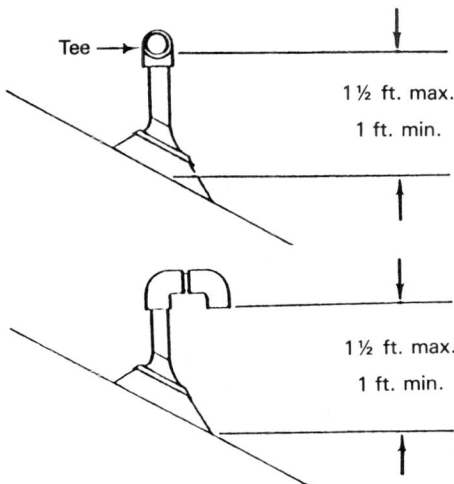

Tee →

1 ½ ft. max.
1 ft. min.

1 ½ ft. max.
1 ft. min.

Figure 6.5 Vent pipe termination recommendations. (Courtesy of The Coleman Company.)

> **Warning:** Failure to follow the following listed requirements and rules can cause unsatisfactory furnace operation or fire, explosion, or asphyxiation.

Special requirements

> **Note:** The first 2 ft of vent, extending from the furnace, must be constructed of CPVC pipe and CPVC fittings. After the first 2 ft, either PVC or CPVC may be used.

1. These furnaces must be vented with 2-in.-diameter (except furnace model 2985, which requires 3-in.-diameter) schedule 40 plastic PVC or CPVC pipe as listed in the specification table. *Do not use any type of metal pipe for venting.*
2. The first 2 ft of vent pipe must be schedule 40 plastic CPVC pipe. A 2-ft-long section of CPVC pipe is supplied with the furnace for this purpose.
3. These furnaces must be vented separately from any other appliance, with the vent running continuously from the furnace to the outside atmosphere. Do not vent these furnaces into a common flue or vent with other appliances, such as water heater, space heater, wood burning stove or fireplace, or clothes drier.
4. Do not vent these furnaces into a masonry chimney or any other all-fuel-type chimney.

Observe the following rules

A. General:
   1. Any horizontal run of the vent must be pitched (a minimum of 1/4-in./ft) upward, never downward—see Figure 6.5.
   2. Joints in the vent pipe must be securely made and any horizontal run of the vent pipe is supported no less than one support every 3 ft to prevent sagging or displacement of the pipe and stressing of the joints and to keep water from restricting the vent pipe.
   3. The vent piping has to be planned to minimize the number of elbows. Short offsets made by joining two elbows together should be avoided if possible. (Two 45° elbows are equal to one 90° elbow in flow resistance.) Approximately, each 5-ft reduction in maximum vent piping length will allow an additional 90° elbow to be used, if needed.
   4. Before attaching the vent pipe, manually spin the vent blower by turning the plastic fan blade on the motor with a small screwdriver. This is to ensure that the blower wheel will spin freely with no interference

between the blower wheel and housing. If there is interference, see the service guide for these furnaces on making the proper adjustments.

5. The 2-ft section of CPVC vent pipe supplied with the furnace must be inserted into the vent blower outlet and, to provide a good mechanical connection to the furnace, securely connected by tightening the bracket supplied with the furnace around the vent pipe (see Figure 6.6).

Figure 6.6 Vent pipe connections to furnace. (Courtesy of The Coleman Company.)

B. Horizontal Vents:
  1. The total run of the vent must not exceed 30 ft with a maximum of two 90° elbows. Keep the vent as short and direct as possible.
  2. When the vent pipe exits the building structure, its exit must be at least 3 ft above any forced air inlet located within 10 ft. Also, the vent must be at least 4 ft below or 4 ft horizontally from or 1 ft above any door, window, or gravity air inlet.
  3. The vent pipe must extend at least 1 ft but no more than 1 and 1/2 ft past the outside wall and must terminate with a 2-in.-diameter schedule 40 or 80 plastic PVC or CPVC tee. The termination should be made so as to prevent possible blockage of the vent with snow and to protect any building materials from degradation by the flue gases (see Figure 6.7).

> **Warning:** The use of a nonapproved terminal could cause fire or asphyxiation.

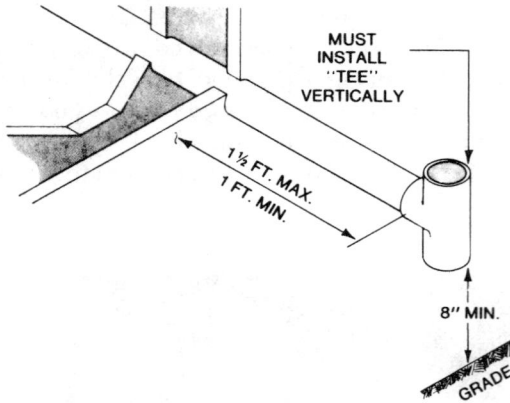

Figure 6.7 Horizontal vent termination. (Courtesy of The Coleman Company.)

Note: For the best results, exit the horizontal vent through any side of the home that does not face the prevailing winter wind.

4. The vent pipe used on model 2985 furnaces must be 3-in.-diameter schedule 40 plastic PVC or CPVC pipe. The pipe that connects to the furnace and that which extends through and past the outside wall must be 2-in.-diameter schedule 40 plastic PVC or CPVC pipe. A suitable 3- to 2-in.-diameter reducer must be used when going from 2- to 3-in.-diameter pipe (see Figure 6.8).

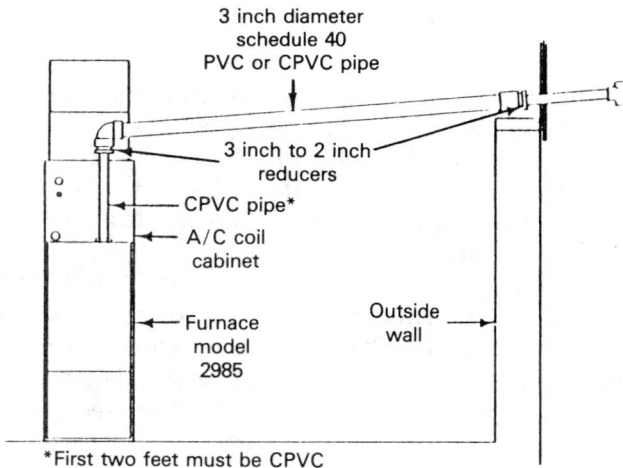

*First two feet must be CPVC

Figure 6.8 Horizontal vent pipe installation. (Courtesy of The Coleman Company.)

C. Vertical Vents:
  1. The total run of the vent must not exceed 30 ft with a maximum of two 90° elbows. Keep the vent as short as possible. The furnace may

be vented vertically by running the vent pipe directly above the vent connection or with a horizontal lateral connected to a vertical riser.

2. Vertical vent risers must not be installed where they will be exposed to the outdoors.

3. The opening where the vent penetrates the roof must be sealed with a plastic flashing.

4. The vent pipe must terminate with a 2-in.- diameter schedule 40 or 80 plastic return bend (two 90° elbows) or tee PVC or CPVC pipe and the opening of the return bend or tee must be at least 1 ft, but no more than 1 and one-half ft above the roof (see Figure 6.5).

Recommendations for cementing PVC or CPVC pipe joints

1. Condition both pipe and fittings to the same temperature conditions.

2. Cut the ends of the pipe square and deburr it. Using a chamfering tool or file, put a 10 to 15° chamfer on the end of the pipe.

3. Using a clean, dry rag, wipe the surfaces to be joined with an all-purpose PVC/CPVC cleaner.

4. Follow the cleaning immediately with the application of an all-purpose PVC/CPVC cement. Apply the cement liberally on the pipe and sparingly on the socket.

5. Quickly push the pipe into the socket with a slight twisting motion until it bottoms. Adjust the alignment of the fitting immediately, before the cement sets.

Condensate Removal

These furnaces are condensing-type appliances and are equipped with a condensate fitting that must be piped to a drain.

Do not install this furnace in an unconditioned space that may have temperatures that may reach 32 °F or below. If the temperature in an unconditioned space ever reaches freezing (32 °F or below), the condensate will freeze in the condensate tube. If this happens, then the condensate will fill the condensing radiator until it blocks the flow of the flue gases, which, in turn, will shut down the furnace. The furnace will not operate until the condensate thaws and unblocks the flow of the flue gases through the condensing radiator.

Special requirements.   The following are some special requirements for these furnaces:

1. The condensate must be removed from the furnace with no less than one-half in. PVC or CPVC pipe (see Figure 6.3). Do not use metal pipe for the removal of condensate from the furnace.

2. The condensate from the furnace may be drained with the condensate from the air conditioning system using a common PVC or CPVC pipe (see Figures 6.9 and 6.10). In this application, the drain pipe must be no less than three-quarter in. i.d.

> **Caution:** The draining of other components not tested in combination with these furnaces may make the equipment in violation of local codes, may create a hazard, and may ruin the equipment and void the warranty.

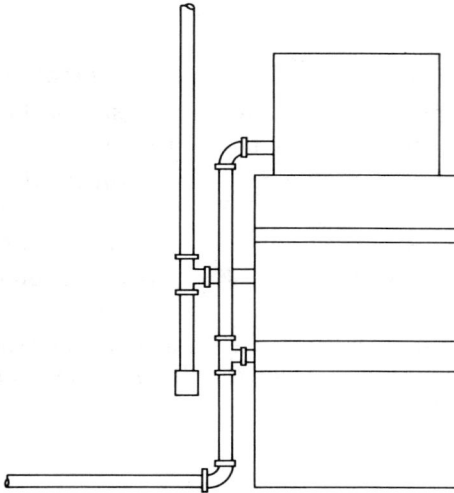

Figure 6.9 Condensate drain installation with air conditioning drain. (Courtesy of The Coleman Company.)

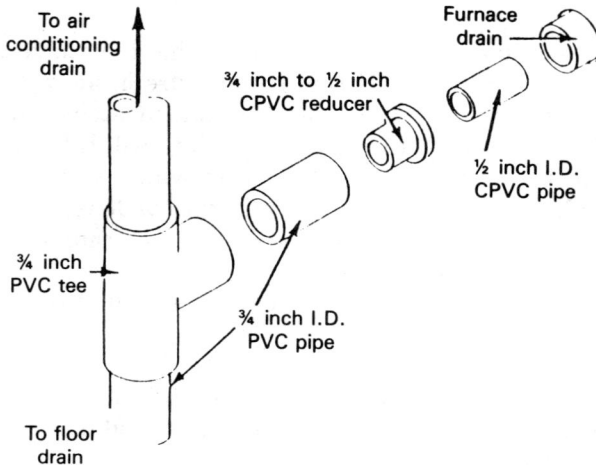

Figure 6.10 Condensate drain connection into air conditioning drain line. (Courtesy of The Coleman Company.)

3. The condensate pipe, from the furnace condensate exit to the drain, must always be lower than the condensate exit point of the furnace.

4. Seal all the condensate pipe joints with a PVC/CPVC cement to ensure that there is no leakage of condensate.

ELECTRICAL WIRING

All internal electric wiring has been made at the factory. Field wiring requires only the connection of line voltage supply wiring and low-voltage thermostat wiring.

Refer to the unit rating plate and specification tables (Table 6.1 of these instructions) for applicable electrical characteristics and requirements.

Service Wiring

Field wiring and electrical grounding of the unit should conform to local codes or in the absence of local codes with the National Electrical Code, ANSI/NFPA 70–1984, if an external electrical source is utilized.

A separate fused circuit from the main electrical panel should serve only the furnace. Connect the power supply leads in the furnace junction box, providing an approved strain relief, as shown in the unit wiring diagram on the unit.

Control Wiring

The thermostat should be installed in accordance with the manufacturer's instructions, furnished with the thermostat, and connected to the unit as shown on the unit wiring diagram. It is recommended that No. 18 AWG wire be used.

If the thermostat has an adjustable heat anticipator, set it to 0.13 A.

Blower Motor Speed Selection

These furnaces are equipped with blowers that have multispeed direct drive motors.

The blower speed selected is dependent upon the design and static pressure loss of the duct system. The duct system external static pressure includes the combined total of the supply and return ducts and any plenum type air conditioning coil if used.

The furnace must be adjusted to operate at or below the maximum external static pressure (in. wc) and within the air temperature rise range as shown on the unit rating plate and the specification table.

Furnace models 29XXB66XX are equipped with 3-speed blower motors.

Furnace models 29XXB76XX are equipped with 4-speed blower motors.

These furnaces are equipped with a blower relay that will change blower speeds automatically when the furnace is properly connected to a heating and cooling-type wall thermostat. The blower motors are factory connected to operate on high or medium (or medium high) speed for heating operation and medium (or medium high) speed for cooling operation. Dependent upon the conditions in a particular installation, the blower speeds may be changed. However, only the high or medium (or medium high) speeds are to be used for heating operation.

Table 6.2 may be used for the selection of the motor speed tap for cooling operation.

> **Danger:** *Shock hazard:* Be sure that the electrical power to the furnace is turned off before changing motor speeds.

TABLE 6.2.   MOTOR SPEED TAP SELECTION FOR COOLING (COURTESY OF THE COLEMAN COMPANY)

Furnace models	Cooling system capacity, Btuh							
	18,000	24,000	30,000	36,000	42,000	48,000	54,000	60,000
2940B66X	Low	Med.	High					
2960B66X		Low	Med.	High				
2970B76X		Low	Low	M. low	M. high	High	High	
2985B76X		Low	Low	M. low	M. high	High	High	
2960B76X				Low	M. low	M. high	High	High

Motor Terminal Identification

In all cases the white wire from the furnace control box is the common circuit and is fitted with a 3/16-in. quick connect terminal. This wire must always be connected to the motor terminal marked white, Common, or 1.

The black wire from the furnace control box is controlled by the furnace fan switch and sequencer (time-delay relay) and is the hot wire for heating operation.

The blue wire is controlled by the blower relay and is the hot wire for cooling operation.

The motor terminals of various motors that may be used are identified by one of the methods shown in Table 6.3.

TABLE 6.3 MOTOR TERMINAL IDENTIFICATION (COURTESY OF THE COLEMAN COMPANY)

Motor speed	Motor terminal identification	
	3-Speed motor	4-Speed motor
High	Common, white or no. 1	Common, white or no. 1
	Hi, black or no. 2	Hi, black or no. 2
Med high		Med hi, blue or no. 3
Medium	Med, blue or no. 3	
Med low		Med lo, yellow or no. 4
Low	Lo, red or no. 4	Lo, red or no. 5

PREOPERATIONAL CHECKS

> **Danger:** *Shock hazard:* Be sure that the electrical power to the furnace is turned off before performing the preoperational checks.

1. Be sure that the furnace is equipped for the type of gas being supplied to the furnace. See the unit rating plate.
2. Make sure that the shipping strap was removed from the blower housing.
3. Manually spin the circulating air blower wheel to ensure that it turns freely and does not strike the blower housing.
4. Was the vent blower wheel checked to ensure that it turned freely before the vent pipe was attached? See venting instructions.
5. Was the gas piping pressure tested and/or purged of air then checked for leaks? See gas piping instructions. Even the smallest leak must be eliminated before attempting to light the furnace.

SEQUENCE OF OPERATION

These furnaces are equipped with an electric hot surface burner ignition system. In response to a call for heat by the room thermostat, the burner is lighted by a glowing ignitor at the beginning of each operation cycle. The burner will continue to operate until the thermostat is satisfied, at which time all burner flame is extinguished. During the off cycle, no gas is used. With the room thermostat set below room temperature and with the electrical power and gas supply to the furnace on, the normal sequence of operation is as follows:

1. When the room temperature falls below the setting of the room thermostat, the thermostat energizes the heating relay.

2. When the heating relay closes, a circuit is made starting the vent blower. A circuit is also made through the normally closed limit switch contacts and the heating relay to the normally open vent air pressure switch contacts.

3. As the vent blower increases in speed, a negative pressure is developed in the vent blower. When sufficient negative pressure has been developed, the contacts of the vent air pressure switch will close and complete the electrical circuit to the electronic ignition module.

4. During the next 40 to 50 seconds, the vent blower will bring fresh air into the heat exchanger and the ignitor will begin to glow. At the end of this period, the ignition module will open the gas valve and energize a safety lockout circuit.

5. When the burner lights, the ignitor acts as a flame probe which checks for the presence of a flame. If a flame is present, the system will monitor it and hold the gas valve open.

6. If the burner fails to light within 6 to 8 seconds after the gas valve opens, the safety lockout circuit in the ignition module will lock off the gas valve and the ignitor. The system will remain in a lockout mode until the room thermostat is set below room temperature. The lockout circuit will then release and setting the room thermostat above room temperature will cause the system to try for ignition again.

7. The lapsed time from the moment that the room thermostat closes, to when the burner lights, may be 50 to 60 seconds. This delay is caused by (1) the time required for the vent blower to come to full speed, (2) the 40 to 50 seconds required for the ignitor to heat up, and (3) the time required for the vent blower to bring fresh air into the heat exchanger.

8. Thirty to forty seconds after the burner has lighted, the normally open contacts of the blower sequencer close and the furnace air circulation blower runs. The fan switch, which is in a parallel circuit with the blower sequencer, will also turn on but at a later time.

> **Note:** If a heating/cooling thermostat is being used and the fan switch is set in the ON (continuous blower) position, the furnace air circulation blower will run at the air conditioning speed. If the wall thermostat calls for heat, the furnace air circulation blower will shut off for 40 to 50 seconds, then the burner will light. Thirty to forty seconds after the burner has lighted, the furnace air circulation blower will begin running again but at the heating speed. There is no pause in the furnace air circulation blower operation when the wall thermostat is satisfied; the furnace air circulation blower just changes over to the cooling (continuous blower) speed.

9. When the room thermostat is satisfied, the circuit to the heating relay is broken and the heating relay contacts return to the normally open position. The circuit to the vent blower, the ignition module, and the blower sequencer is broken and the burner is extinguished. The contacts of the blower sequencer return to the normally open position within 30 seconds after the burner extinguishes. The fan switch remains on until it senses a fall in temperature to a safe limit, then opens the circuit to the furnace air circulation blower.

OPERATION CHECKS

The gas valve is multifunctional, with two operating valves in line, a pressure regulator, and a manual gas cock (see Figures 6.11–6.13). The pressure regulator is factory set to provide an operating manifold pressure of 3.5 in. wc on models equipped for natural gas and 10 in. wc on models equipped for LP gas.

The pressure regulator is a limited adjustment type. Refer to the section Minor Input Adjustment under Furnace Input Capacity.

Lighting Instructions

This furnace is equipped with an automatic hot surface ignition system that lights the main burner each time the thermostat calls for heat.

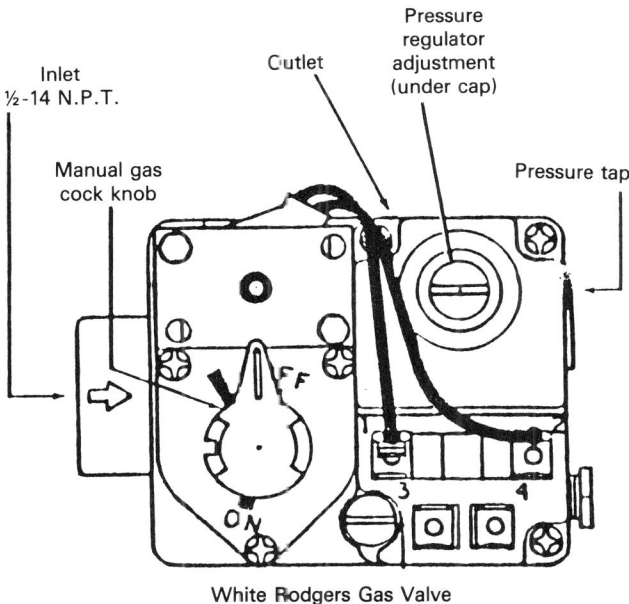

White Rodgers Gas Valve

Figure 6.11 White Rodgers gas valve connections and components. (Courtesy of The Coleman Company.)

Inlet
½-14 N.P.T.

Manual gas
cock knob

Outlet

Pressure
regulator
adjustment
(under cap)

Pressure tap

PUSH-TURN ON
OFF

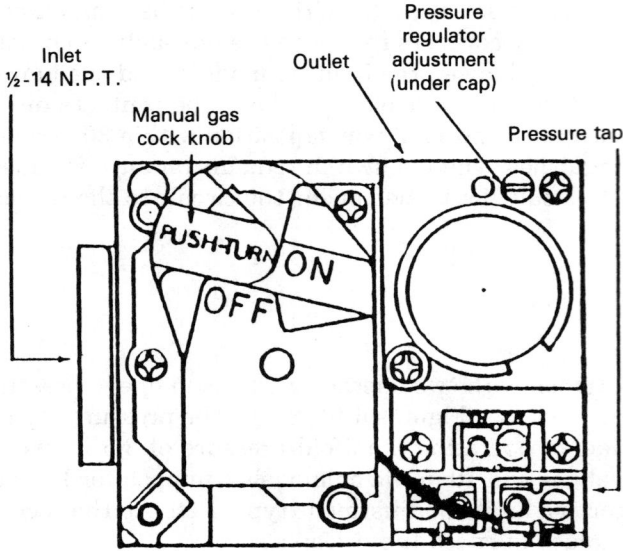

Honeywell Gas Valve

Figure 6.12 Honeywell gas valve connections and components. (Courtesy of The Coleman Company.)

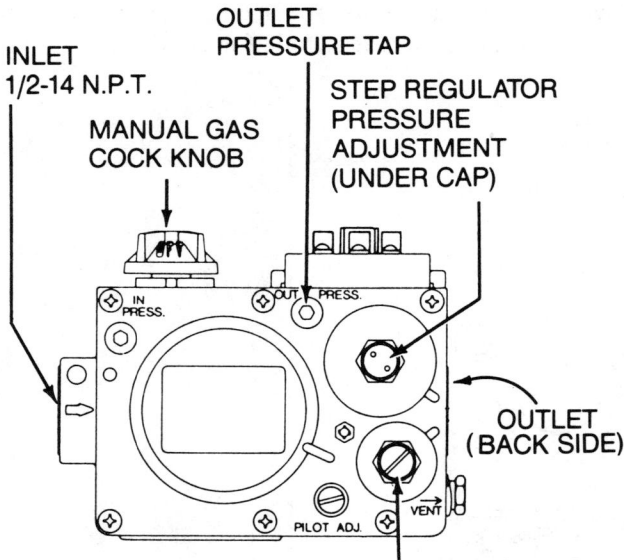

INLET
1/2-14 N.P.T.

OUTLET
PRESSURE TAP

MANUAL GAS
COCK KNOB

STEP REGULATOR
PRESSURE
ADJUSTMENT
(UNDER CAP)

IN
PRESS.

OUT. PRESS.

OUTLET
(BACK SIDE)

PILOT ADJ.

VENT

MAIN GAS PRESSURE
REGULATOR ADJUSTMENT
(UNDER CAP)

ROBERTSHAW GAS VALVE

Figure 6.13 Robertshaw gas valve connections and components. (Courtesy of The Coleman Company.)

This furnace cannot be lighted with a match.

> **Warning:** Failure to follow these instructions may result in an explosion and possible damage to the furnace and injury to the operator.

1. Set the room thermostat to the lowest setting or OFF.
2. Turn the knob on the gas control valve to OFF. The White Rodgers gas valve has a built-in stop that prevents the knob from being turned directly to OFF. At this stop the knob must be depressed and then turned to OFF (see Figure 6.11).
3. Wait 5 minutes.
4. Turn the knob on the gas control valve to ON.
5. Set the room thermostat to the desired setting, above room temperature. The burner will then light, which may take 50 to 60 seconds.
6. If the burner fails to light, go to a complete shutdown and determine the cause for the failure to light. Be sure that the gas supply to the furnace is on and the supply piping has been purged of air. Be sure that the electric power is on.
7. For a complete shutdown, set the thermostat to the lowest setting or OFF and turn the knob on the gas control valve to OFF.

Burner Adjustment

After lighting the furnace, allow the furnace to operate for approximately 15 minutes and then adjust the burner primary air as follows:

1. Loosen the air adjustment rod locking screw (see Figure 6.14).
2. Close the primary air shutter by pulling on the adjustment rod until yellow tips appear in the flame at the end of the burner (see Figure 6.15).
3. Slowly open the primary air shutter by pushing the adjustment rod in until the yellow tips at the end of the burner disappear; then push the adjustment rod in another 1/8 in.
4. Secure the primary air adjustment rod by tightening the locking screw against it.

Furnace Input Capacity

The maximum Btuh input capacity for each model is shown on the unit rating plate and in the specification tables. This input must not be exceeded.

The input shown may be used in geographical areas where the elevation is from 0 to 2000 ft above sea level. In areas above 2000 ft (high altitude), the

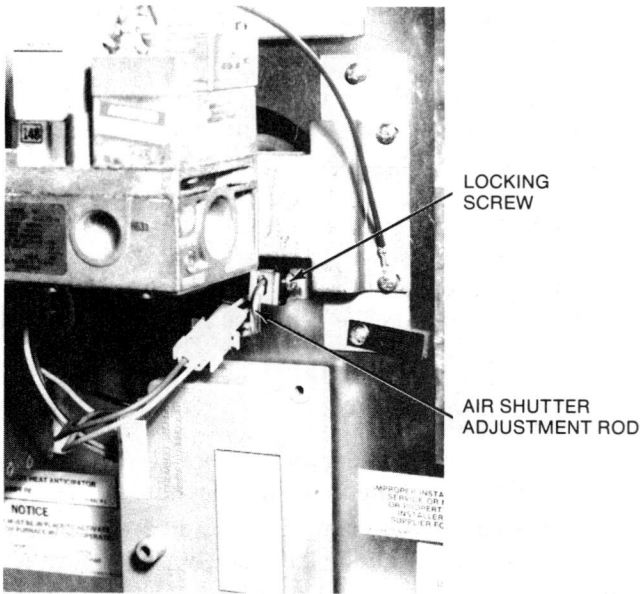

LOCKING
SCREW

AIR SHUTTER
ADJUSTMENT ROD

Figure 6.14 Burner air shutter adjustment. (Courtesy of The Coleman Company.)

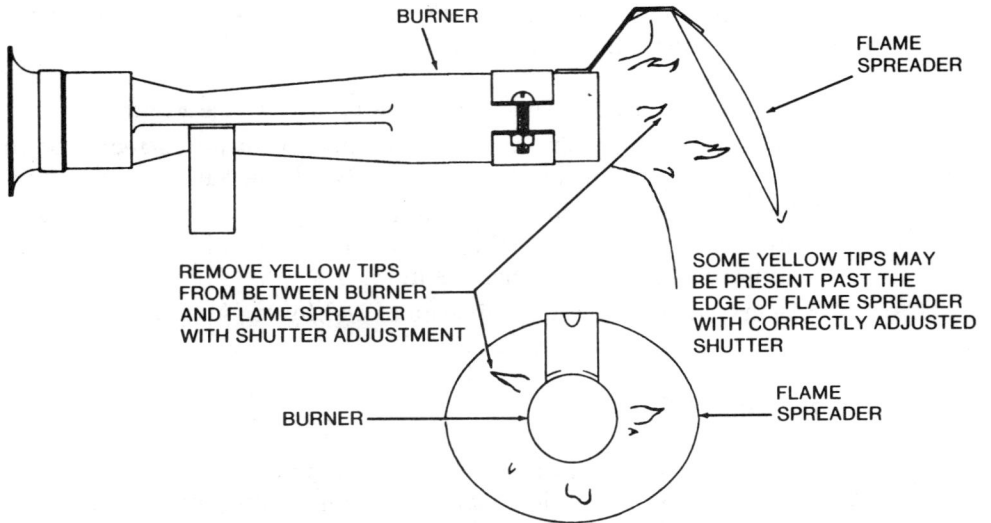

BURNER

FLAME
SPREADER

REMOVE YELLOW TIPS
FROM BETWEEN BURNER
AND FLAME SPREADER
WITH SHUTTER ADJUSTMENT

SOME YELLOW TIPS MAY
BE PRESENT PAST THE
EDGE OF FLAME SPREADER
WITH CORRECTLY ADJUSTED
SHUTTER

BURNER

FLAME
SPREADER

Figure 6.15 Burner flame adjustment. (Courtesy of The Coleman Company.)

furnace BTU input must be reduced 4% for each 1000 ft of elevation above sea level. The Btu input depends on the calorific value of the gas (Btu/ft³), orifice size, and manifold pressure. Coleman orifice sizes are based on calorific values of 1050 Btu/ft³ for natural gas and 2500 Btu/ft³ for LP gas (propane). The orifice size supplied with the furnace should provide satisfactory input capacity for installations in most areas, except at high altitude.

Minor input adjustment. The input may be adjusted slightly by adjusting the pressure regulator in the gas control valve to change the manifold pressure.

To adjust the pressure regulator, remove the cover screw on top of the valve (see Figures 6.11–6.13). Turn the adjusting screw counterclockwise to decrease the pressure and clockwise to increase the pressure. In no case should the final manifold pressure vary more than ±0.3 in. wc from the specified regulator pressure settings: 3.5 in. wc for natural gas and 10 in. wc for LP gas.

> **Danger:** Never attempt to modify this furnace: Fire, explosion, or asphyxiation may result. If a malfunction is apparent, contact a qualified service agency, and/or gas utility for assistance.

Determining gas input rate. Where the gas is metered, the input rate may be determined by the following method:

Contact the gas supplier, public utility company, or LP gas distributor to obtain the calorific value of the gas being used. When checking the input rate, any other gas burning appliances connected to the same meter should be completely off. The furnace should be allowed to operate for approximately 15 minutes before attempting to check the gas flow rate.

To check the flow rate, observe the 1-ft³ dial on the gas meter and determine the number of seconds required for the dial to make one revolution (seconds to flow 1 ft³).

To determine the number of seconds required for the flow of one cubic foot of gas, use the following formula:

$$\frac{\text{(Btu content)}}{\text{Calorific value of gas} \times 3600}$$
$$\text{Furnace Btuh input}$$

Example:

1000 Btu gas, furnace input 100 000 Btu/hour.

$$\text{Seconds for 1 ft}^3 = \frac{1000 \times 3600}{100\,000} = 36 \text{ seconds}$$

If, when checking the meter, the 1-ft^3 dial makes a complete revolution in less time than was calculated that it should, the furnace is overfired and will need to be derated. If, in clocking the meter, it takes more time than was calculated for the meter to make one revolution, the furnace is underfired.

The orifice size must be changed to correct an overfired or underfired condition. If it is determined that a different orifice is needed, please contact the distributor for assistance in selecting the correct replacement.

Balance the System

It is recommended that the air distribution system be balanced, using in-line duct dampers if installed, to provide for satisfactory air delivery room to room. It is recommended that dampers in the registers not be used for balancing the system.

Cycle furnace for correct operation.   After all parts of the installation have been completed (furnace installation, vent installation, and duct system installation), the furnace should be checked for normal operation prior to the time that the system will be used.

1. Place the heating system in the operating mode by following the lighting instructions outlined previously in these instructions and on the furnace rating plate.
2. Cycle the furnace off and on a number of times, allowing the furnace to light and the circulating air blower to come on. After the thermostat is turned off allow time for both circulating air blower and the vent blower to stop before the next operating cycle.
3. While the unit is in operation, check to determine that the circulating air blower is operating smoothly with no undue vibration or noise. Check to determine that the vent blower is operating smoothly with no noise or vibration. Check to ensure that the main burner has been adjusted and the main burner flame is normal.
4. Turn off the furnace at the thermostat and turn off the electric power supply to the furnace. Disconnect the black wire from the vent blower and isolate the wire terminal to prevent any contact with the furnace casing or ground to eliminate the possibility of a shock hazard. Turn on the electric power to the furnace and turn on the thermostat. Under this condition the vent blower should not run and the furnace should not light. Turn off the thermostat and the electrical power and then reconnect the black wire to the vent blower.

Important: There are improper operating conditions that could cause water to form in the furnace where it is not designed to collect. These conditions may occur when:

1. The furnace is operated with return air temperatures below 55 °F. Examples would be during new construction in place of construction heaters to
   try to warm the interior of the structures, or owners using night set back
   thermostats set below 55 °F. Coleman recommends not operating this
   furnace with return air temperatures below 55 °F. One exception would
   be for the short time needed to check out the furnace after installation.
2. The furnace is underfired while operating because the Btu content of the
   fuel being used is low.
3. The burner is improperly adjusted, causing the furnace to short cycle
   while operating.

If more than one of these conditions exist and water collects other than
where the flue gases were designed to condense, the furnace could become
blocked by the water and not operate.

Refer to the service guide for more details concerning these conditions.

SERVICE MAINTENANCE

A program of periodic inspection and preventative maintenance can help ensure trouble-free operation, provide more efficient operation, and extend the
life of the furnace.

> Danger: To avoid the possibility of an electric shock, turn off the electrical
> power supply to the furnace before any disassembly for inspection or
> maintenance work is performed.

Homeowner's Maintenance

The following are the recommended maintenance steps for this furnace:

Circulating air blower assembly.   Use the following steps to maintain the
circulating air blower assembly:

1. Check to ensure that the impeller (blower wheel) turns freely.
2. Check for buildup of foreign material on the blades. Clean off any accumulation with a stiff brush or by scraping with a suitable tool.

Take care not to damage the blades or move any balancing weights that may be on the blades.

3. Motor Lubrication: If the motor is provided with oil ports, lubricate the motor with SAE 20 nondetergent-type oil. Slowly add 10 to 15 drops to each oil port. *Do not over oil.* Motors that are not provided with oil ports require no lubrication. Vent and Condensate Tubing: Inspect the vinyl tubing connecting the vent pressure switch and the vent blower and the vinyl tubing used for the condensate removal. If there are any signs of hardening, cracking, or deterioration, the tubing must be replaced. Ordinary rubber hose is not satisfactory and must not be used.

Service Inspection

Service inspection and repair of the furnace, and any related equipment that may be installed in conjunction with the furnace, shall be done only by persons who are qualified and experienced with this type of equipment. Some state and municipal code authorities require that persons servicing this type of equipment be licensed. Persons who are not qualified or licensed shall not attempt to service this equipment.

Coleman recommends that the furnace be inspected by a qualified service company, such as a Coleman dealer, shortly before the beginning of each heating season. The furnace should be inspected prior to the beginning of each air conditioning season, if air conditioning equipment is installed.

When required, adjustments should be made to ensure the operational performance of the furnace is in accordance with the specifications shown on the unit rating plate and in these installation instructions.

Before removing components or assemblies for inspection, carefully observe how these components are installed, electrical connections made, and the attachment means used. All components and assemblies must be reinstalled in the same manner and position as they were originally.

Warning: Never attempt to modify this furnace or repair damages or defective components. Such action could cause unsafe operation, fire, explosion, or asphyxiation.

Damaged or defective components must be replaced. Only original equipment components, or substitute components authorized by the Coleman Company, Inc., may be used in the repair of this furnace.

Condensate tubing.   The condensate tubing water trap at the bottom of the furnace vestibule should be periodically cleaned of accumulations of particles and dirt to allow free flow of the condensate.

Burner assembly. The burner should be inspected for any accumulation of corrosion (rust) in the burner port. The burner should be blown out with the pressure side of a vacuum cleaner. Take care not to damage the ignition assembly.

> **Note**: The ignitor assembly is very fragile and should be handled with care. A cracked or broken ignitor will not function properly and must be replaced.

> **Caution:** The ignitor operates on line voltage. Turn off the electrical power before servicing to prevent a shock hazard which could damage the equipment or cause personal injury.

Heat Exchanger

The heat exchanger should be inspected for possible material corrosion, scaling, damage, or deposits of soot.

1. Remove:
   a. The gas valve mounting bracket.
   b. The burner assembly.
2. Place a suitable light through the combustion air tube and into the heat exchanger drum. Using a mirror, observe the heat exchanger inner surfaces and look for any deposit of black soot or any apparent scaling on the metal surfaces. The normal color appearance of the heat exchanger may have a red color in some areas.
3. If soot deposits are found in the heat exchanger drum and the condensate tubes show blackness from sooty condensate water, the heat exchanger must be replaced and the cause of the soot determined and necessary corrective action taken.
4. If any light sooting is found, it and any debris from the bottom of the heat exchanger must be removed using a suitable vacuum cleaner. If sooting is noticed, its cause must be determined and necessary corrective action taken.

REPLACEMENT PARTS

Should it be necessary to replace any component parts, these may be obtained through a Coleman dealer, who is experienced and can be of assistance, or information on the nearest distributor can be obtained directly from the Coleman Company, Inc., 3050 N. St. Francis, Wichita, KS 67219, phone (316) 832–6448.

> **Caution:** Only genuine Coleman replacement parts should be used. Substitute parts should not be used because they may not be the same in operational and safety characteristics.

OVERALL OPERATION

Following major service or replacement of functional parts, it is recommended that the furnace be operated in the various modes to ensure that performance is normal and the control components are functioning properly.

THE COLEMAN T.H.E. MODEL "B" GAS FURNACE UNIT WIRING DIAGRAMS

90 SERIES COLEMAN GAS FORCED AIR FURNACE

USE ONLY 115 VAC 60 HZ 1 PH
LESS THAN 12 AMPS   MAX. OVERCURRENT PROTECTION 15 AMPS

FACTORY INTERNAL WIRING SHOWN SOLID
IF ANY OF THE ORIGINAL WIRE SUPPLIED WITH THIS UNIT MUST BE RE-
PLACED, IT MUST BE REPLACED WITH TYPE 105°C THERMOPLASTIC
OR ITS EQUIVALENT.
*ORANGE AND TAN WIRES INDICATED: 200°C
†WHITE RODGERS HEAT THERMOSTAT: COLEMAN MODEL 7670-3751
COOLING SUB BASE: COLEMAN MODEL 7670-3701
WHEN OTHER MODEL THERMOSTAT IS USED, REFER TO MFG. DIAGRAM
WITH THERMOSTAT AND SUB BASE FOR CONNECTION
WHEN USING 2-WIRE HEATING ONLY THERMOSTAT, CONNECT TO RED
AND WHITE WIRES.
‡DO NOT USE THESE BLOWER TAPS FOR HEATING SPEEDS.

DANGER: SHOCK HAZARD
TURN OFF ELECTRICAL POWER BEFORE SERVICING FURNACE TO PRE-
VENT SHOCK HAZARD WHICH COULD DAMAGE EQUIPMENT OR CAUSE
PERSONAL INJURY.

LADDER WIRING DIAGRAM

ABBREVIATIONS

APS = Air Pressure Switch
BDSS = Blower Door Safety Switch
BM = Blower Motor
BR = Blower Relay
CAP = Capacitor
EA = Electrode Assembly
FS = Fan Switch
GV = Gas Valve
H/C T = Heat/Cool Thermostat
HR = Heating Relay
HVC = High Voltage Coil
IM = Ignition Module
LS = Limit Switch
SEQ = Sequencer
VM = Vent Motor
XFMR = Transformer

MODELS
2940, 2960, 2970,
AND 2985

THE COLEMAN COMPANY, INC.
Wichita, Kansas 67201

1972A225 (11-83) P.I.

ELECTRICAL WIRING

All internal electrical wiring has been made at the factory. Field wiring requires only the connection of Line Voltage supply wiring and Low Voltage thermostat wiring.

Refer to the unit rating plate and specification tables found in these instructions for applicable electrical characteristics and requirements.

Service Wiring

Field Wiring and electrical grounding of the unit should conform to local codes or in the absence of local codes with the National Electrical Code, ANSI/NFPA 70-1984, if an external electrical source is utilized

A separate fused circuit from the main electrical panel should serve only the furnace. Connect the power supply leads in the furnace junction box, providing an approved strain relief, as shown in the unit wiring diagram on the unit and in these instructions.

Control Wiring

Thermostat should be installed in accordance with the manufacturer's instructions, furnished with the thermostat, and make connections to the unit as shown in the unit wiring diagram. It is recommended 18 AWG wire be used.

If the thermostat has an adjustable heat anticipator, set it to .13 ampere.

Blower Motor Speed Selection

These furnaces are equipped with blowers which have multi-speed direct drive motors.

The blower speed selected is dependent upon the design and static pressure loss of the duct system. The duct system external static pressure includes the combined total of the supply and return ducts and any plenum type air conditioning coil if employed.

The furnace must be adjusted to operate at or below the maximum external static (in. W.C.) and within the air temperature rise range as shown on the unit rating plate and in the specification table.

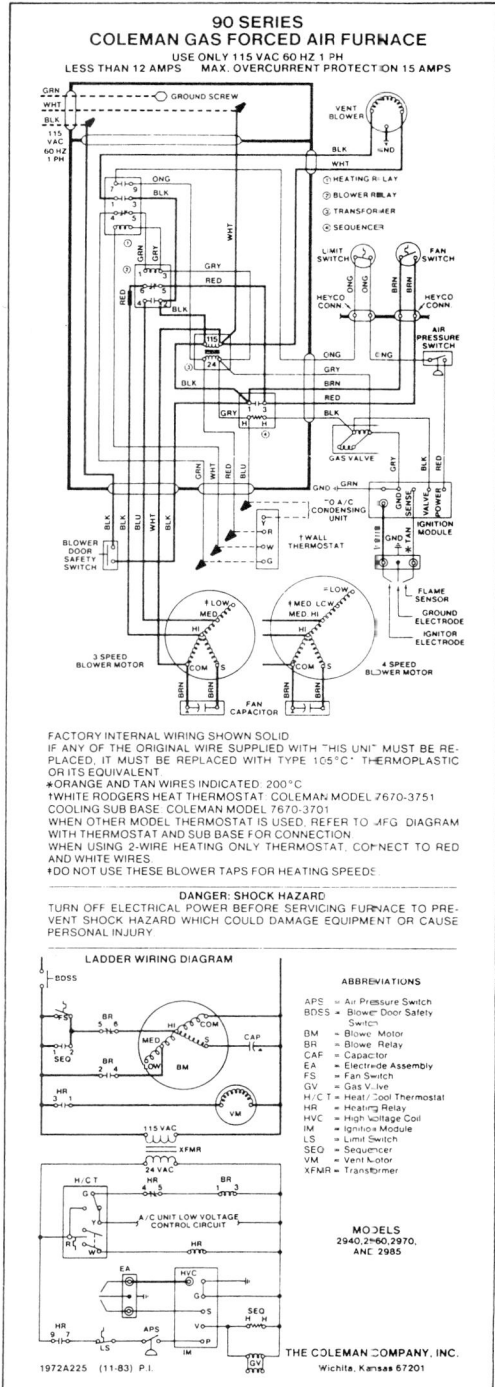

COLEMAN GAS FORCED AIR FURNACE
USE ONLY 115 VAC 60 HZ 1 PH
LESS THAN 12 AMPS   MAX. OVERCURRENT PROTECTION 15 AMPS

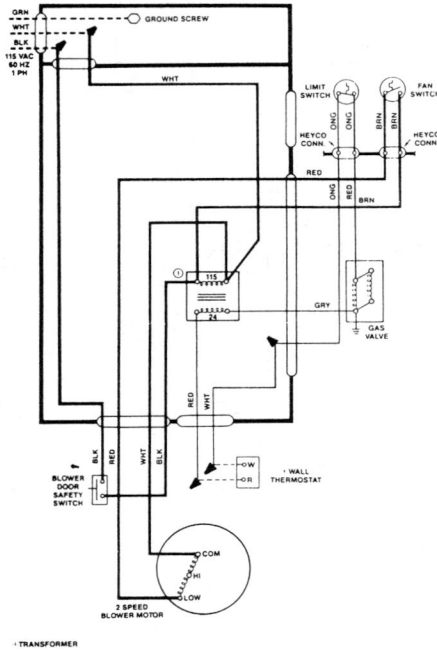

†TRANSFORMER

FACTORY INTERNAL WIRING SHOWN SOLID.
IF ANY OF THE ORIGINAL WIRE SUPPLIED WITH THIS UNIT MUST BE REPLACED,
IT MUST BE REPLACED WITH TYPE 105 °C THERMOPLASTIC OR ITS EQUIV-
ALENT.
†WHITE RODGERS HEAT THERMOSTAT: COLEMAN MODEL 7670-3751
WHEN OTHER MODEL THERMOSTAT IS USED, REFER TO MFG. DIAGRAM WITH
THERMOSTAT AND SUB BASE FOR CONNECTION.

DANGER: SHOCK HAZARD
TURN OFF ELECTRICAL POWER BEFORE SERVICING FURNACE TO PREVENT
SHOCK HAZARD WHICH COULD DAMAGE EQUIPMENT OR CAUSE PERSONAL
INJURY.

LADDER WIRING DIAGRAM

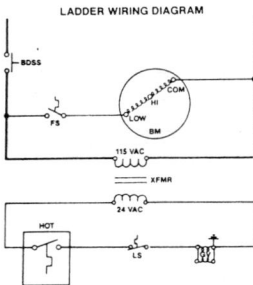

ABBREVIATIONS

BDSS - Blower Door Safety
         Switch
BM   - Blower Motor
FS   - Fan Switch
GV   - Gas Valve
HOT - Heating Only Thermostat
LS   - Limit Switch
XFMR - Transformer

MODELS
2735-646
2755-646

1951-851 Rev. 1 (10-86) P.I.

THE COLEMAN COMPANY, INC.
Wichita, Kansas 67201

COLEMAN GAS FORCED AIR FURNACE
USE ONLY 115 VAC 60 HZ 1 PH
LESS THAN 12 AMPS   MAX. OVERCURRENT PROTECTION 15 AMPS

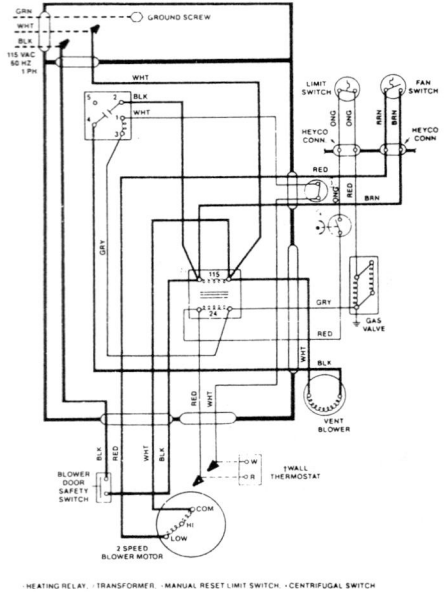

·HEATING RELAY, ·TRANSFORMER, ·MANUAL RESET LIMIT SWITCH, ·CENTRIFUGAL SWITCH

FACTORY INTERNAL WIRING SHOWN SOLID
IF ANY OF THE ORIGINAL WIRE SUPPLIED WITH THIS UNIT MUST BE REPLACED,
IT MUST BE REPLACED WITH TYPE 105 C THERMOPLASTIC OR ITS EQUIV-
ALENT.
†WHITE RODGERS HEAT THERMOSTAT: COLEMAN MODEL 7670-3751
WHEN OTHER MODEL THERMOSTAT IS USED, REFER TO MFG. DIAGRAM WITH
THERMOSTAT AND SUB BASE FOR CONNECTION.

DANGER: SHOCK HAZARD
TURN OFF ELECTRICAL POWER BEFORE SERVICING FURNACE TO PREVENT
SHOCK HAZARD WHICH COULD DAMAGE EQUIPMENT OR CAUSE PERSONAL
INJURY

LADDER WIRING DIAGRAM

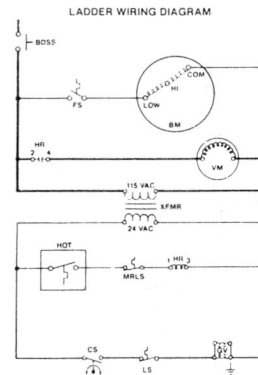

ABBREVIATIONS

BDSS - Blower Door Safety
         Switch
BM   - Blower Motor
CS   - Centrifugal Switch
FS   - Fan Switch
GV   - Gas Valve
HOT - Heating Only Thermostat
HR   - Heating Relay
LS   - Limit Switch
MRLS - Manual Reset Limit Switch
VM   - Vent Motor
XFMR - Transformer

MODELS
2775-746
2790-746

1951A852 (10-86) P.I.

THE COLEMAN COMPANY, INC.
Wichita, Kansas 67201

COLEMAN GAS FORCED AIR FURNACE

COLEMAN GAS FORCED AIR FURNACE

USE ONLY 115 VAC 60 HZ 1 PH
LESS THAN 12 AMPS   MAX. OVERCURRENT PROTECTION 15 AMPS

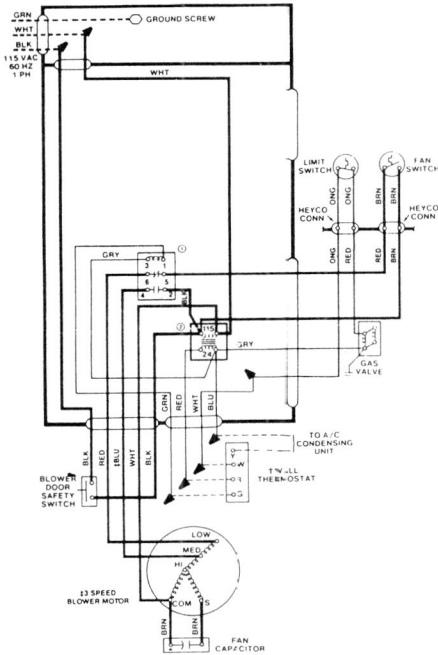

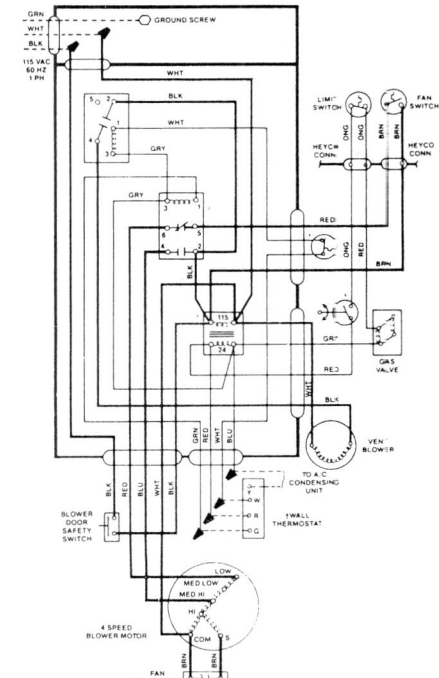

① BLOWER RELAY. ② TRANSFORMER

FACTORY INTERNAL WIRING SHOWN SOLID.
IF ANY OF THE ORIGINAL WIRE SUPPLIED WITH THIS UNIT MUST BE RE-
PLACED, IT MUST BE REPLACED WITH TYPE 105°C THERMOPLASTIC
OR ITS EQUIVALENT
†WHITE RODGERS HEAT THERMOSTAT: COLEMAN MODEL 7670-3751
COOLING SUB BASE: COLEMAN MODEL 7670-3701
WHEN OTHER MODEL THERMOSTAT IS USED, REFER TO MFG. DIAGRAM
WITH THERMOSTAT AND SUB BASE FOR CONNECTION.
WHEN USING 2-WIRE HEATING ONLY THERMOSTAT. CONNECT TO RED
AND WHITE WIRES.
‡MODEL 2755-726 HAS A FOUR SPEED BLOWER MOTOR AND HAS THE BLUE
WIRE ATTACHED TO THE MEDIUM HIGH POSITION.

DANGER: SHOCK HAZARD
TURN OFF ELECTRICAL POWER BEFORE SERVICING FURNACE TO PRE-
VENT SHOCK HAZARD WHICH COULD DAMAGE EQUIPMENT OR CAUSE
PERSONAL INJURY

LADDER WIRING DIAGRAM

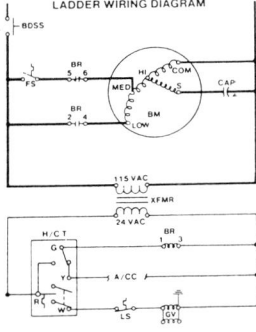

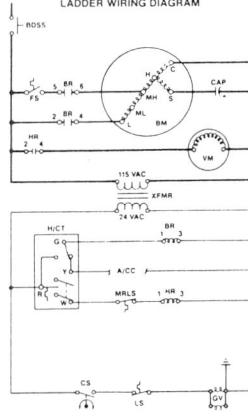

ABBREVIATIONS	
A/CC	A/C Unit Low Voltage Control Circuit
BDSS	Blower Door Safety Switch
BM	Blower Motor
BR	Blower Relay
CAP	Capacitor
FS	Fan Switch
GV	Gas Valve
H/C T	Heat/Cool Thermostat
LS	Limit Switch
XFMR	Transformer

MODELS
2735-626
2745-626
2755-626
‡2755-726

THE COLEMAN COMPANY, INC.
Wichita, Kansas 67201

1951-853 Rev. 1 (10-86) P.I.

COLEMAN GAS FORCED AIR FURNACE

USE ONLY 115 VAC 60 HZ 1 PH
LESS THAN 12 AMPS   MAX. OVERCURRENT PROTECTION 15 AMPS

• HEATING RELAY • BLOWER RELAY • TRANSFORMER • MANUAL RESET LIMIT SWITCH • CENTRIFUGAL SWITCH

FACTORY INTERNAL WIRING SHOWN SOLID
IF ANY OF THE ORIGINAL WIRE SUPPLIED WITH THIS UNIT MUST BE REPLACED.
IT MUST BE REPLACED WITH TYPE 105 C THERMOPLASTIC OR ITS EQUIV-
ALENT
†WHITE RODGERS HEAT THERMOSTAT: COLEMAN MODEL 7670-3751
COOLING SUB BASE: COLEMAN MODEL 7670-3701
WHEN OTHER MODEL THERMOSTAT IS USED, REFER TO MFG. DIAGRAM WITH
THERMOSTAT AND SUB BASE FOR CONNECTION
WHEN USING 2-WIRE HEATING ONLY THERMOSTAT. CONNECT TO RED AND
WHITE WIRES.

DANGER: SHOCK HAZARD
TURN OFF ELECTRICAL POWER BEFORE SERVICING FURNACE TO PREVENT
SHOCK HAZARD WHICH COULD DAMAGE EQUIPMENT OR CAUSE PERSONAL
INJURY

LADDER WIRING DIAGRAM

ABBREVIATIONS	
A/CC	A/C Unit Low Voltage Control Circuit
BDSS	Blower Door Safety Switch
BM	Blower Motor
BR	Blower Relay
CAP	Capacitor
CS	Centrifugal Switch
FS	Fan Switch
GV	Gas Valve
H/C T	Heat/Cool Thermostat
HR	Heating Relay
LS	Limit Switch
MRLS	Manual Reset Limit Switch
VM	Vent Motor
XFMR	Transformer

MODELS
2775-726
2790-726

THE COLEMAN COMPANY, INC.
Wichita, Kansas 67201

1951-854 REV. 1 (9-86) P.I.

7

Electronic Ignition Systems

Electronic ignition systems are designed to conserve energy by shutting off the pilot burner gas when there is no call for heat from the thermostat. These controls save from 3 to 5% in most installations.

COMPONENTS AND PARTS DESCRIPTION

An intermittent ignition device (IID) includes the following components:

1. Ignition Control: This component houses the electronic circuitry used to control the safety, sequencing, and operation of the heating system.
2. Gas Valve: The gas valve electronically controls both the pilot and the main burner gas supply. It may also include a gas pressure regulator and a manual shutoff valve.
3. Electrode Pilot Burner: The electrode pilot burner incorporates an electrode that lights the pilot gas on a call for heat from the temperature control.
4. Flame Sensor: The flame sensor detects the presence of the pilot flame and acts as a key part of the sensing circuit.

SEQUENCE OF OPERATION

The following is the sequence of operation of electronic ignition systems:

1. When the temperature control calls for heat, the spark transformer in the ignition control and the pilot gas valve are automatically energized.
2. The spark lights the pilot gas on each operating cycle.
3. The flame sensor proves the presence of the flame. The ignition control then shuts off the spark transformer. At the same time, the main burner gas valve is energized. (Some models permit the spark to continue for a short period of time after the main burner gas is ignited.) On 100% lockout models, a shutdown of the entire system will occur if the pilot gas does not light within some fixed period of time (usually 30 seconds).
4. The main burner gas lights and the system continues the normal operating cycle.
5. When the temperature control is satisfied, the main burner and pilot gas valves are deenergized, shutting off all gas flow to the burners.

PRINCIPLES OF OPERATION

Basically there are two different types of electronic ignition systems in use today. One type uses a spark to ignite the burner gas and the other type uses a hot surface to ignite the burner gas.

The spark ignition systems can be further divided into two more types: (1) the intermittent ignition device (IID), and (2) the direct spark ignition (DSI) system. The IID is the most popular spark ignition system used on residential units. It is used to light the pilot burner gas, which, in turn, lights the main burner gas. In these installations, the pilot flame must be proven before the main burner valve is energized.

The direct spark ignition system uses a spark to light the main burner gas directly without the use of a pilot burner. These are also popular on residential units and in commercial applications that require energy savings and automatic ignition for the main burner gas. Some manufacturers use a spark plug similar to the one used in an automobile and others make use of a spark electrode located directly over the main burner.

Sensing Methods

When a standing pilot is used, heat is used to operate the thermocouple to energize the pilot safety circuit. In installations where the IID is used, flame conduction or rectification is used. To aid in understanding the principles of flame conduction, we must first have a thorough understanding of the structure of a gas flame (see Figure 7.1).

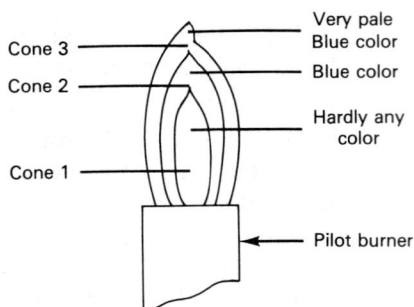

Figure 7.1 Structure of a gas flame.

When the flame is properly adjusted with the proper air–gas ratio there are three cones or zones present:

1. In cone 1, or the inner cone, the flame will not burn because there is an excess of fuel to air.
2. Cone 2, or the intermediate cone, surrounds the inner cone. This area has a slightly different color of blue and is where the proper combustion occurs. This cone takes a part of its combustion air from the surrounding air, or the secondary air.
3. Cone 3, also known as the outer cone or envelope, contains an excessive amount of air which it gets from the surrounding secondary air. This envelope is also a different color of blue from the other two cones.

Cone 2, or the intermediate cone, is the place that we are most concerned with. Since this is the place where the best combustion occurs, it is a prime location for the sensing probe of the electronic ignition device.

In actuality, a flame is but a series of small, controlled, explosions that cause the immediate area to become ionized. The ionization of this area causes the area to become electrically conductive (see Figure 7.2). This conductive flame is often thought of as being an electrical switch, which is located between the flame sensor and the pilot burner tip. If there is not a flame between the sensing probe and the pilot burner tip, the switch is open. When the flame makes contact with both the pilot burner tip and the sensing probe, the switch is closed. In Figure 7.3 the flame is used to conduct an ac signal. In this case, both of the probes have approximately the same amount of area exposed to the flame. However, this is not a safety device signal because it does not identify the type of current conducted by the flame. This type of sytem could mistake an electrical short and allow the main gas valve to be opened without a

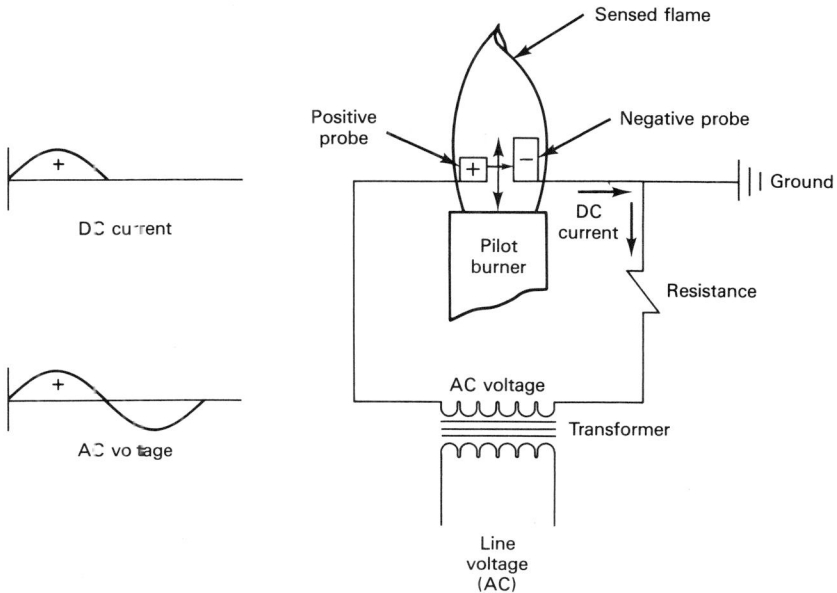

Figure 7.2 Principle of flame rectification.

means of ignition. To prevent this from happening, a difference in the type of flame must be noted. This difference is known as flame rectification (see Figure 7.4).

Flame Rectification

When flame rectification is used, the flame and probes are used in a similar manner, with one important exception: The area of one of the probes exposed to the flame must have a greater area than the other probe.

In Figure 7.4 the flame is again used to conduct an ac signal. Both of the probes are in contact with the flame. In this case the probe with the greatest area attracts the most free electrons and becomes electrically negative. In this condition, the direction of the flow of electrons and, therefore, the current is from the positive probe to the negative probe. It should be noted that the ac voltage sine wave has not been changed, but the negative portion of the current sine wave is not present. Thus, the ac current has been changed to pulsating dc current. The result is flame rectification.

To make this principle useful in IIDs, a pilot and flame sensor must be used in place of the two probes (see Figure 7.4b). After ignition of the pilot, a microampere dc current flow is conducted through the flame, from the flame sensor (the positive probe) to the pilot burner tip (the negative probe). Acting

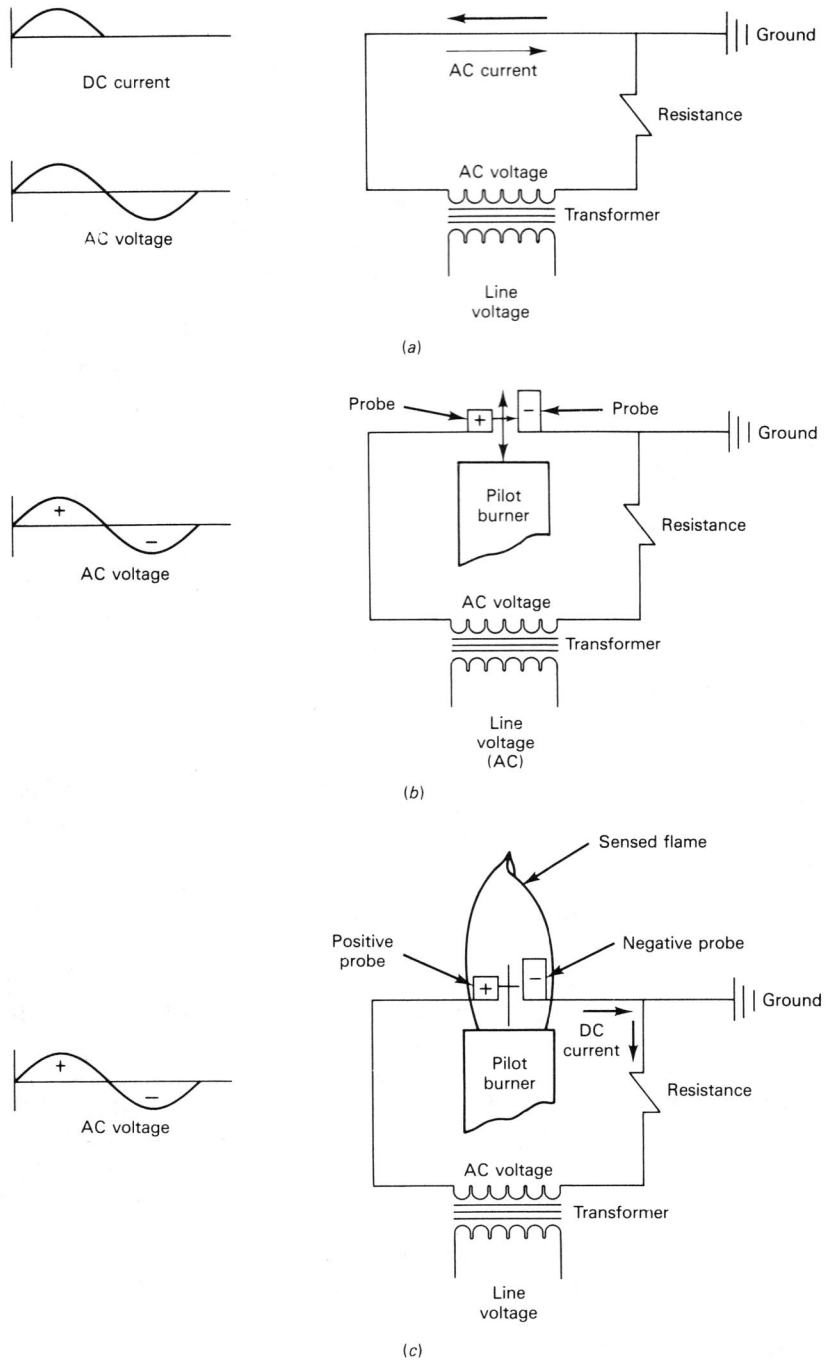

Figure 7.3 (a) Normal ac circuit: (b) flame out, switch is open, no current is flowing; (c) pilot is lit, flame is sensed, and current is flowing.

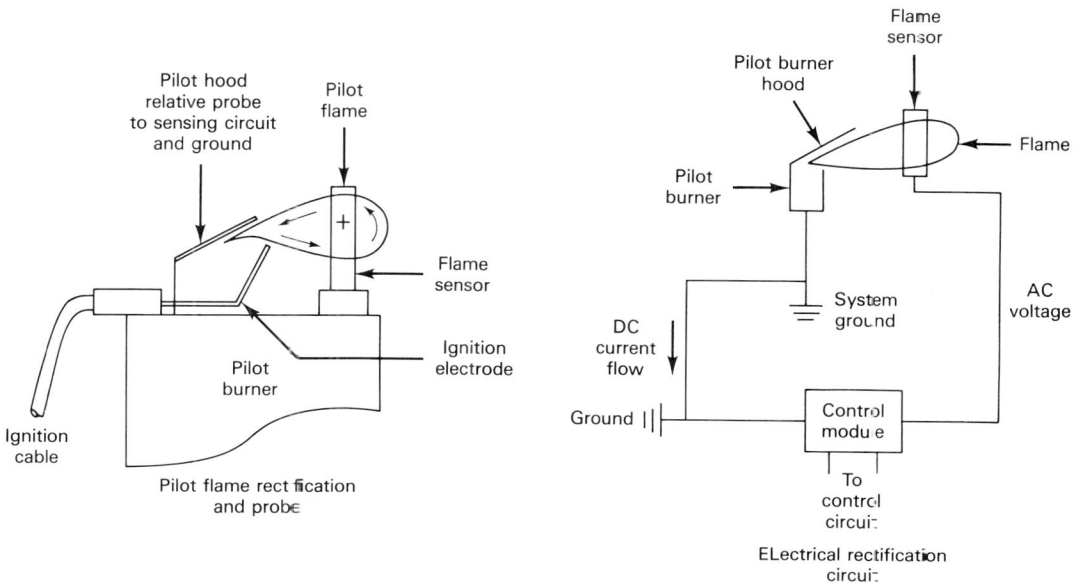

Figure 7.4 Electrical rectification by use of pilot flame.

as the negative probe, the pilot burner tip completes the electrical circuit to ground. The IID system uses this dc current flow to energize a relay, which, in turn, energizes the main burner gas valve.

The following conditions have a direct bearing on every IID application.

Voltage. The supply voltage to the ignition controls should be within the following ranges:

120-VAC controls: 102 to 132 VAC
24-VAC controls: 21 to 26.5 VAC

24-VAC systems should use transformers that will provide adequate power under maximum load conditions.

Gas Pressure. The inlet gas pressure should be a minimum of 1 in. wc above the equipment manufacturer's recommended gas manifold pressure. The inlet gas pressure must never be allowed to fall below the equipment manufacturer's recommended minimum inlet gas pressure.

The maximum inlet gas pressure for natural gas should be limited to 10.5 in. wc. On LP applications, the inlet gas pressure should be limited to a maximum of 14 in. wc.

Temperature.   Electronic ignition controls should never be exposed to temperatures greater than 150 °F or less than −40 °F.

Pilot Application

The pilot and flame sensor application is the most critical aspect of the IID application.

The pilot flame must make contact with the pilot burner tip and completely surround the sensor probe. A microammeter is necessary to verify that the proper amount of current is maintained through the pilot flame. If the proper amount of current is not maintained for the equipment in use, the unit will not operate as it was designed. This could be indicated by rapid short cycling of the main burner flame or no main burner flame on. Flame rectification ignition systems respond in less than 0.8 seconds to a loss of flame signal. Thus, any deflection of the pilot flame away from the flame sensor, or the pilot burner tip, could result in rapid cycling (chattering) of the main burner gas valve, or prevent the main burner from coming on at all.

Other conditions that could cause the failure of the main burner to come on or rapid chattering of the main burner valve are (1) a pilot flame that is too small, or (2) the gas pressure being too low for proper pilot flame impingement on the flame sensor, but the main burner gas valve will not be energized. It is also posible for drafts or unusual air currents to deflect the pilot flame away from the flame sensor. Deflection of the pilot flame may also be caused by main burner ignition concussion or rollout of the main burner flame.

An additional point to be considered is the condition of the pilot flame. If the pilot flame is hard and blowing, the grounding area of the pilot is reduced to a point so that the necessary current flow is not being maintained, and a shutdown of the system will result.

The positioning of the flame sensor is also critical in the pilot application. Positioning of the flame sensor should be such that it will be in contact with the second or combustion area of the pilot flame. Passing the flame sensor through the inner cone of the pilot flame is not a recommended procedure. For this reason, a short flame sensor may provide a superior signal over a longer one. The final determination of the sensor location (length) is best determined by the use of a microammeter.

CHECKING OUT INTERMITTENT PILOT SYSTEMS

This section is presented with permission of Honeywell Incorporated.
To check out the gas control system use the following steps:

1. At the initial installation of the appliance.
2. As a part of regular maintenance procedures. Maintenance intervals are determined by the application as indicated in the following section.

3. As the first step in troubleshooting.
4. Any time that work is done on the system.

APPLICATIONS

Electronic ignition systems are used on a wide variety of central heating equipment and on heating appliances such as commercial cookers, agricultural equipment, industrial heating equipment, and outdoor pool heaters. Two important reasons for choosing electronic ignition are:

1. To save fuel.
2. To avoid nuisance shutdowns on equipment mounted where the pilot can be easily blown out or fouled.

Fuel savings result from eliminating the standing pilot. This is the motivation behind building codes in some areas that prohibit standing pilots in new construction.

Furnaces located in crawl spaces, in attics, or on rooftops are excellent candidates for electronic ignition because the pilot is more apt to be blown out and because it is often hard to reach. Furnaces and heating appliances in locations where the air is dusty or greasy often use electronic ignition to avoid problems associated with clogging the pilot orifice.

Some of these applications may make heavy demands on electronic ignition systems also, because of moisture, corrosive chemicals, dust, or excessive heat in the environment. In these situations, special steps may be required to prevent nuisance shutdowns and premature control failure. These situations require Honeywell Residential Division Engineering review; contact a Honeywell sales representative for assistance. With the exception of intermittent pilot retrofit, do not apply electronic ignition systems in the field.

> **Warning:** Fire or explosion hazard. Can cause fire or explosion with property damage, injury, or loss of life.

1. If you smell gas or suspect a gas leak, turn off the gas at the normal service valve and evacuate the house. Do not try to light any appliance. Do not touch any electrical switch or the telephone in the building until you are sure that no spilled gas remains.
2. A gas leak test, described in steps 1 and 5 below, must be done on the initial installation and any time that work is done involving the gas piping.

Step 1: Perform a Visual Inspection.
    1. With the electric power off, make sure that all wiring connections are clean and tight.
    2. Turn on the electric power to the appliance and the ignition module.
    3. Open the manual shutoff valve in the gas line to the appliance.
    4. Do a gas leak test ahead of the gas control if the gas piping has been disturbed.

> **Gas Leak Test:** Paint the pipe joints with a rich soap-and-water solution. Bubbles indicate a gas leak. Tighten the joints to stop the leak.

Step 2: Review the normal operating sequence and timing summary (see Figures 7.5 and 7.6 and Table 7.1).

Step 3; To Reset the Module.
    1. Turn the thermostat to its lowest setting.
    2. Wait one minute.
        As you do steps 4 and 5, watch for points where the unit operation deviates from normal. Refer to the Troubleshooting Chart to correct the problem.

Step 4: Check the Safety Lockout Operation.
    1. Turn the gas supply off.
    2. Set the thermostat or controller above room temperature to call for heat.

TABLE 7.1   INTERMITTENT PILOT MODULE TIMING SUMMARY (COURTESY OF HONEYWELL)

Function	S86A, E	S86B, F S90A	S86C, G	S86D, H S90B	S860
Prepurge	—	—	—	—	30 sec.
Trial for ignition (Pilot gas and spark)	Continuous	Continuous	Ends with lockout	Ends with lockout[b]	Ends with lockout
Safety lockout timing	—	—	15 or 90 sec.[a]	15 or 90 sec.[a, c]	15 or 90 sec.[a]
Flame failure response[d]	0.8 sec. max.	0.8 sec. max.	0.8 sec. max.	0.8 sec. max.	0.8 sec. max.

[a]Lockout timing is stamped on module.

[b]S86H, S90B built before 6/87 continue spark after lockout until system is reset.

[c]S90B not available with 15 sec. safety lockout timing.

[d]Shutdown may be delayed several seconds if flame is lost immediately after being proved.

Start

| ① | Thermostat (controller calls for heat |

Trial for
ignition

| ② | Safe start check, must not be any flame simulating condition. |

Power interruption, system shuts off, restarts when power is restored.

Pilot flame failure, second main operator closes, module starts trial for ignition.

| ③ | Spark generator powered first valve (pilot) operator opens |

If flame simulation condition present, system fails to start.

Main burner
operation

④ Pilot burner operation

Pilot burner does not light

Pilot burner lights module senses flame current.   or

Model	Response
S86A,B,E,F S90A	Ignition spark continues, pilot valve remains open until stat is reset
S86H, S90B	After 15 or 90 sec[a], pilot valve closes, spark continues until stat is reset

[a] Lockout timing is stamped on module.

| ⑤ | If flame current sensed • Spark generator off, • Second valve operator (main) opens |

| ⑥ | Main burner operation, module monitors pilot flame current. |

End

| ⑦ | Thermostat (controller) satisfied - valves close, pilot and main burners are off. |

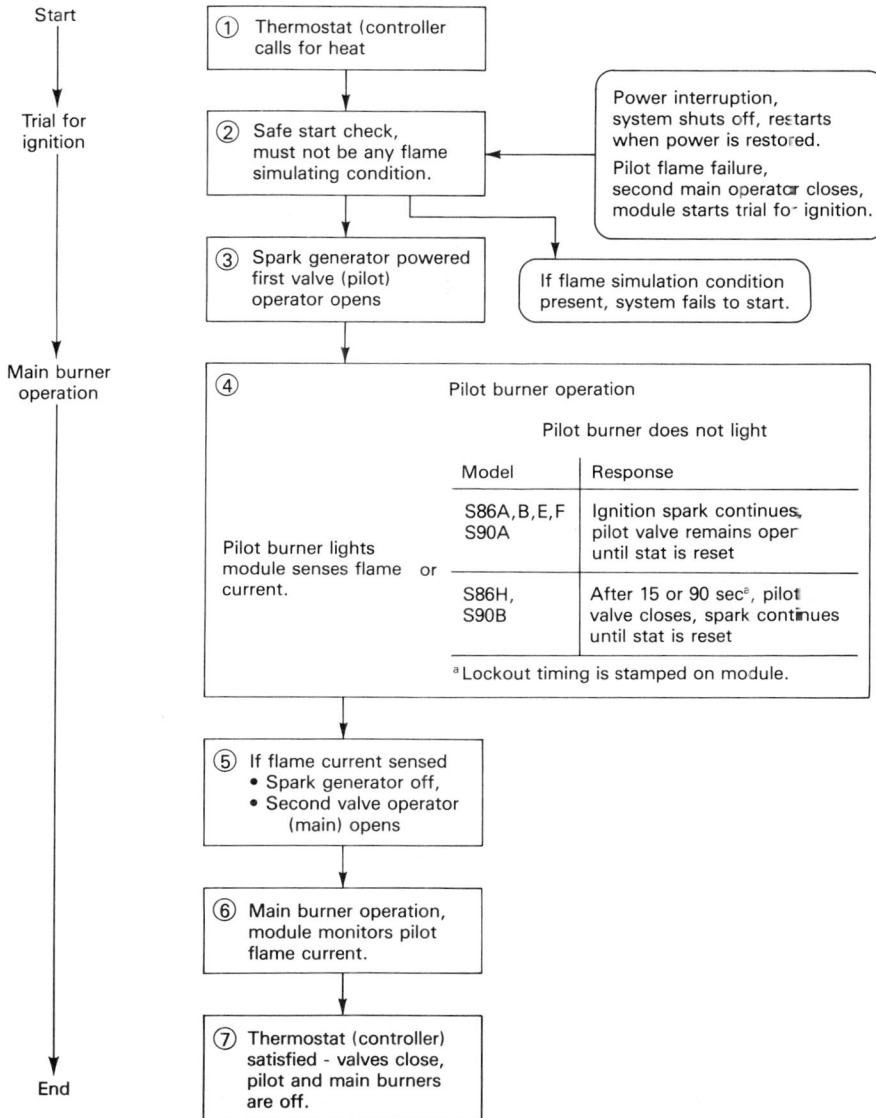

Figure 7.5 S86 and S90 normal operating sequence. (Courtesy of Honeywell.)

3. Watch for sparking at the pilot burner either immediately or following a prepurge (see the timing summary, Table 7.1).
4. On models with timed ignition, time the length of spark operation (see the timing summary, Table 7.1).
5. On models that continue to spark after lockout, check the time before

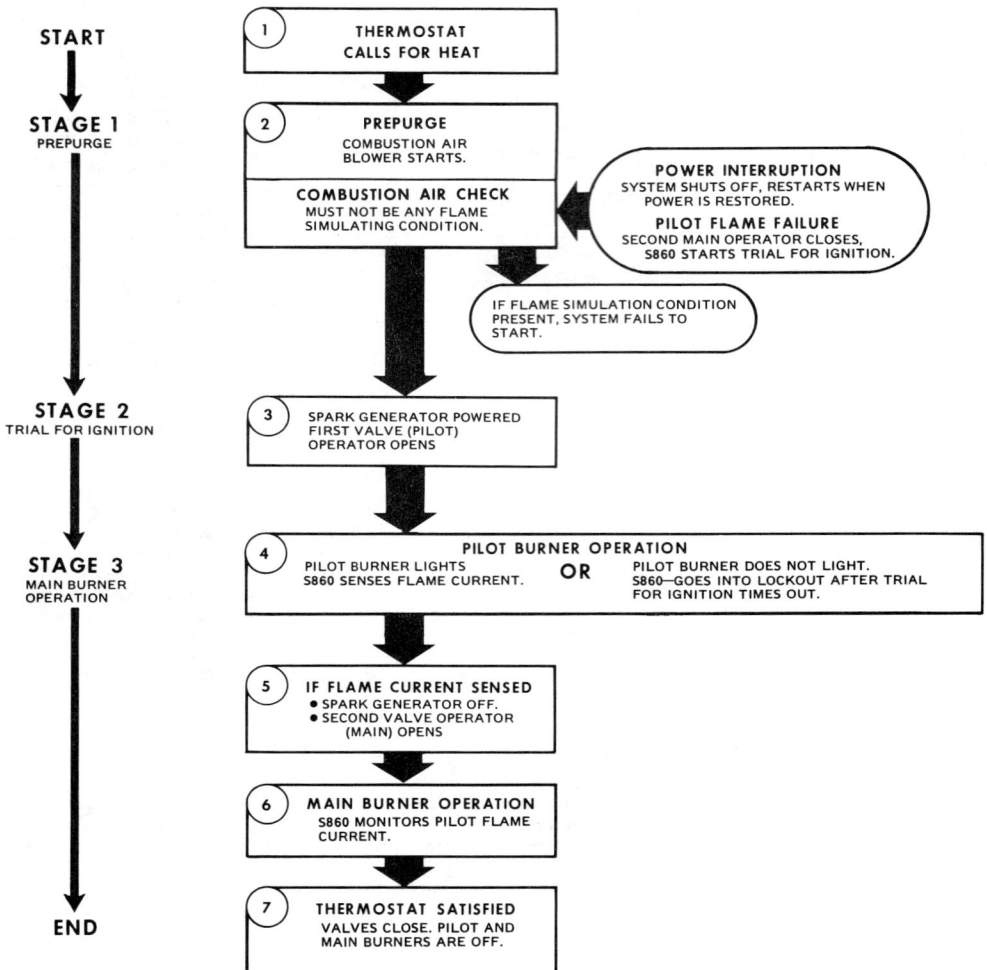

START	① THERMOSTAT CALLS FOR HEAT
STAGE 1 PREPURGE	② PREPURGE COMBUSTION AIR BLOWER STARTS.
	COMBUSTION AIR CHECK MUST NOT BE ANY FLAME SIMULATING CONDITION.

POWER INTERRUPTION
SYSTEM SHUTS OFF, RESTARTS WHEN POWER IS RESTORED.
PILOT FLAME FAILURE
SECOND MAIN OPERATOR CLOSES, S860 STARTS TRIAL FOR IGNITION.

IF FLAME SIMULATION CONDITION PRESENT, SYSTEM FAILS TO START.

STAGE 2 TRIAL FOR IGNITION	③ SPARK GENERATOR POWERED FIRST VALVE (PILOT) OPERATOR OPENS
STAGE 3 MAIN BURNER OPERATION	④ PILOT BURNER OPERATION PILOT BURNER LIGHTS S860 SENSES FLAME CURRENT. **OR** PILOT BURNER DOES NOT LIGHT. S860—GOES INTO LOCKOUT AFTER TRIAL FOR IGNITION TIMES OUT.
	⑤ IF FLAME CURRENT SENSED ● SPARK GENERATOR OFF. ● SECOND VALVE OPERATOR (MAIN) OPENS
	⑥ MAIN BURNER OPERATION S860 MONITORS PILOT FLAME CURRENT.
END	⑦ THERMOSTAT SATISFIED VALVES CLOSE. PILOT AND MAIN BURNERS ARE OFF.

11,191A

Figure 7.6 S860 normal operating sequence. (Courtesy of Honeywell.)

you hear a click from the gas control. A click indicates that the gas control has closed in a safety lockout.
6. Open the manual gas cock and make sure that no gas is flowing to either the pilot or the main burner.
7. Set the thermostat below room temperature and wait one minute before continuing.

Step 5: Check Normal Operation.
1. Set the thermostat or controller above room temperature to call for heat.

2. Make sure that the pilot lights smoothly when the gas reaches the pilot burner.
3. Make sure that the main burner lights smoothly without flashback.
4. Make sure that the burner flame operates smoothly without floating, lifting, or flame rollout to the furnace vestibule or heat buildup in the vestibule.
5. If the gas line has been disturbed, complete a gas leak test.

> **Gas Leak Test:** Paint the gas control gasket edges and all pipe connections downstream of the gas control, including the pilot tubing connections, with a rich soap-and-water solution. Bubbles indicate gas leaks. Tighten the joints and screws or replace the component to stop the gas leak.

6. Turn the thermostat or controller below room temperature. Make sure that the main burner and pilot flames go out.

TROUBLESHOOTING INTERMITTENT PILOT SYSTEMS

Follow the appropriate troubleshooting guide to pinpoint the cause of the problem (see Figures 7.7 and 7.8). If troubleshooting indicates an ignition problem, see Ignition System Checks below to isolate and correct the problem. Following the troubleshooting procedure, perform the checkout procedure above again to be sure that the system is operating normally.

Ignition system checks. Use the following steps when making ignition system checks:

Step 1: Check the Ignition Cable.
　　　　Make sure that:
1. The ignition cable does not touch any metal surfaces.
2. The ignition cable is no more than 36 in. (0.9 m) long.
3. The connections to the stud terminal and the ignitor–sensor are clean and tight.
4. The ignition cable provides good electrical continuity.

Step 2: Check the Ignition System Grounding. Nuisance shutdowns are often caused by a poor or erratic ground.
1. A common ground, usually supplied by the pilot burner bracket, is required for the module and pilot burner/ignitor sensor.
    A. Check for a good metal-to-metal contact between the pilot burner bracket and the main burner.
    B. Check the ground lead from the GND terminal on the module to the pilot burner. Make sure that the connections are clean and

START

TURN THERMOSTAT (CONTROLLER) TO CALL FOR HEAT.

TURN GAS SUPPLY OFF

POWER TO MODULE (24 Vac NOMINAL)

NOTE: BEFORE TROUBLESHOOTING, FAMILIARIZE YOURSELF WITH THE STARTUP AND CHECKOUT PROCEDURE.

NO → CHECK LINE VOLTAGE POWER, LOW VOLTAGE TRANSFORMER, LIMIT CONTROLLER, THERMOSTAT (CONTROLLER) AND WIRING. ALSO, CHECK AIR PROVING SWITCH ON COMBUSTION AIR BLOWER SYSTEM (IF USED).

YES ↓

SPARK ACROSS IGNITER/SENSOR GAP

NO → PULL IGNITION LEAD AND CHECK SPARK AT IGN. STUD ON MODULE / SPARK OKAY?

NO →
- CHECK FUSE (IF INCLUDED). REPLACE IF NECESSARY.
- ON OTHER MODELS, REPLACE MODULE.

YES ↓

- CHECK IGNITION CABLE, GROUND WIRING, CERAMIC INSULATOR AND GAP, AND CORRECT.
- CHECK BOOT OF THE IGNITION CABLE FOR SIGNS OF MELTING OR BUCKLING. TAKE PROTECTIVE ACTION TO SHIELD CABLE AND BOOT FROM EXCESSIVE TEMPERATURES.

YES ↓

TURN GAS SUPPLY ON

PILOT BURNER LIGHTS?

NO →
- CHECK THAT ALL MANUAL GAS VALVES ARE OPEN, SUPPLY TUBING AND PRESSURES ARE GOOD, AND PILOT BURNER ORIFICE IS NOT BLOCKED.
- CHECK ELECTRICAL CONNECTIONS BETWEEN MODULE AND PILOT OPERATOR ON GAS CONTROL.
- WITHIN TRIAL FOR IGNITION TIMING, CHECK FOR VOLTAGE ACROSS PV-MV/PV TERMINALS ON MODULE:

MODULE	APPROXIMATE VOLTAGE
S86C	4-5 Vdc
S86A,B,D,E,F,H	24 Vac
S86G	8 Vac
S90A,B	24 Vac

- IF VOLTAGE IS OK, REPLACE GAS CONTROL. IF NOT, REPLACE MODULE.

YES ↓

SPARK STOPS WHEN PILOT IS LIT?

NO →
NOTE: IF S86C,D,G,H AND S90B GO INTO LOCKOUT, RESET SYSTEM.
- CHECK CONTINUITY OF IGNITION CABLE AND GROUND WIRE.
- CHECK THAT PILOT FLAME COVERS ELECTRODE.
- ADJUST FLAME ROD SO FLAME CURRENT READING ON MICROAMMETER IS AT LEAST:

S86A,B,D,E,F,H	1.0 µA
S86C,G	1.2 µA
S90A,B	1.0 µA

- IF PROBLEM PERSISTS, REPLACE MODULE.

YES ↓

MAIN BURNER LIGHTS?

NO →
- CHECK FOR VOLTAGE ACROSS MV-MV/PV TERMINALS ON MODULE:

MODULE	APPROXIMATE VOLTAGE
S86A THRU SERIES 3	10 Vdc
S86A SERIES 4 AND LATER	24 Vac
S86C,G	10 Vdc
S86B,D,E,F,H	24 Vac
S90A,B	24 Vac

- IF VOLTAGE IS OK, REPLACE GAS CONTROL. IF NOT, REPLACE MODULE.
- CHECK ELECTRICAL CONNECTIONS BETWEEN MODULE AND GAS CONTROL. IF OKAY, REPLACE GAS CONTROL OR GAS CONTROL OPERATOR.

YES ↓

SYSTEM RUNS UNTIL CALL FOR HEAT ENDS?

NO →
NOTE: IF S86C,D,G,H AND S90B GO INTO LOCKOUT, RESET SYSTEM.
- CHECK CONTINUITY OF IGNITION CABLE AND GROUND WIRE. NOTE: IF GROUND IS POOR OR ERRATIC, SHUTDOWNS MAY OCCUR OCCASIONALLY EVEN THOUGH OPERATION IS NORMAL AT THE TIME OF CHECKOUT.
- IF CHECKS ARE OKAY, REPLACE MODULE.

YES ↓

CALL FOR HEAT ENDS

SYSTEM SHUTS OFF?

NO →
- CHECK FOR PROPER THERMOSTAT (CONTROLLER) OPERATION.
- REMOVE MV LEAD AT MODULE; IF VALVE CLOSES, RECHECK TEMPERATURE CONTROLLER AND WIRING; IF NOT, REPLACE GAS CONTROL.

YES ↓

TROUBLESHOOTING ENDS — REPEAT PROCEDURE UNTIL TROUBLEFREE OPERATION IS OBTAINED.

M 106

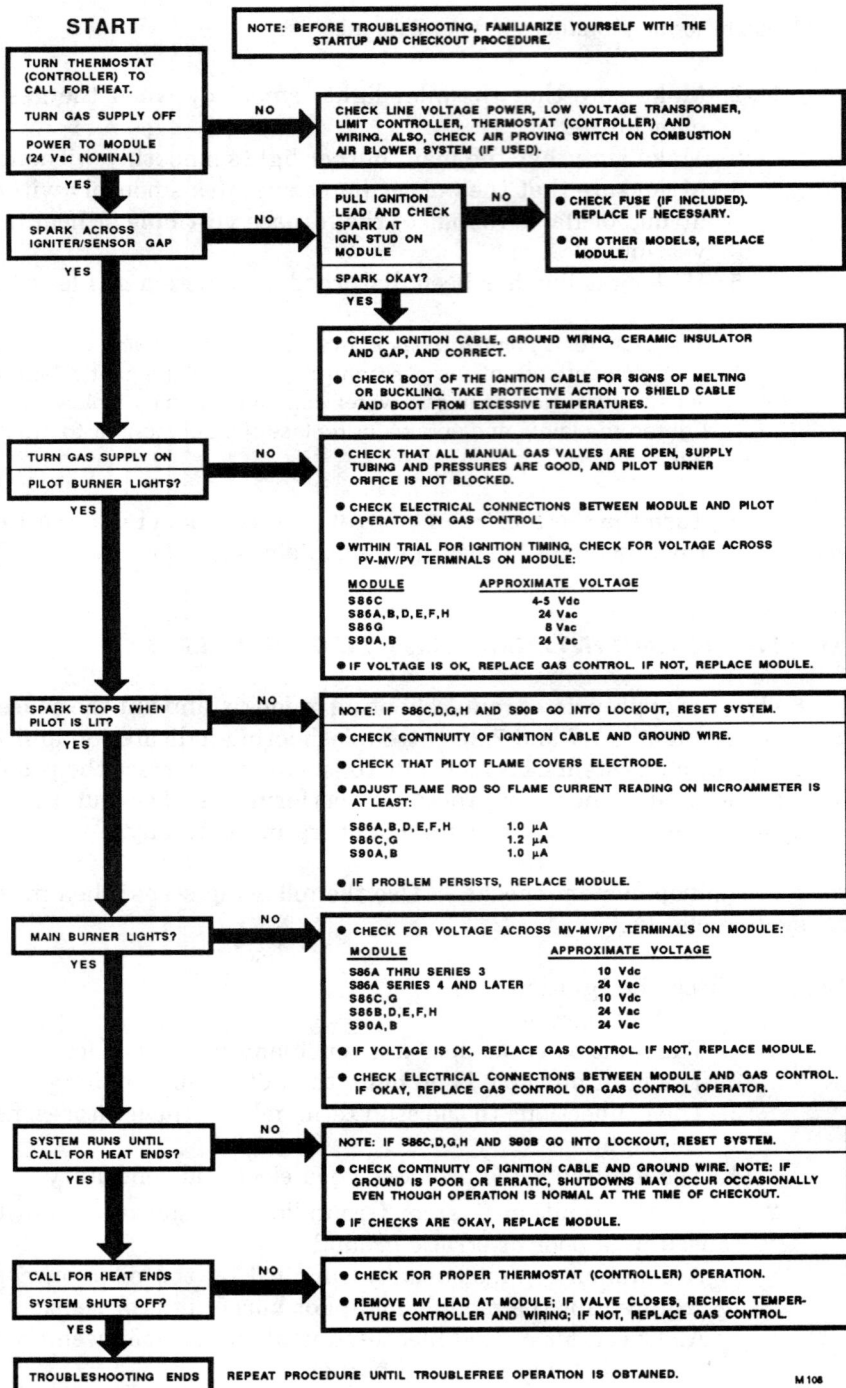

Figure 7.7 S86 and S90 troubleshooting guide. (Courtesy of Honeywell.)

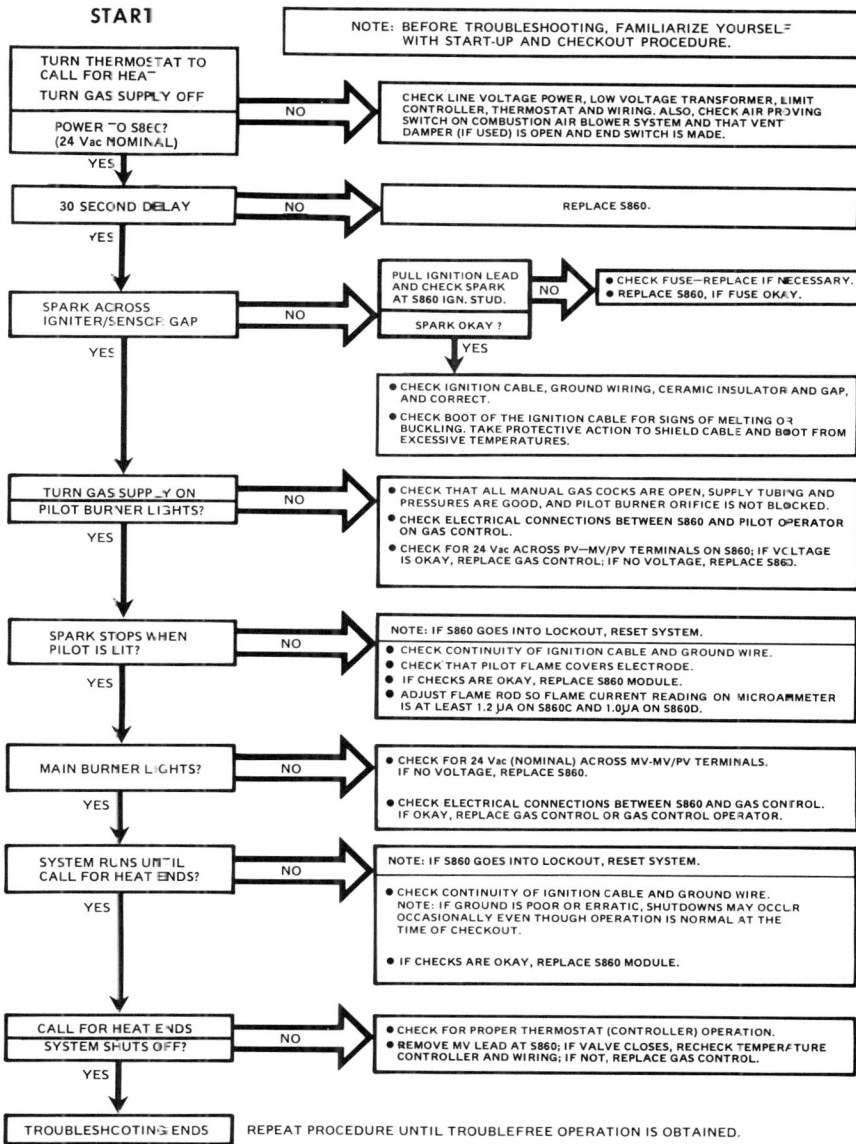

START

NOTE: BEFORE TROUBLESHOOTING, FAMILIARIZE YOURSELF WITH START-UP AND CHECKOUT PROCEDURE.

TURN THERMOSTAT TO CALL FOR HEAT
TURN GAS SUPPLY OFF
POWER TO S860? (24 Vac NOMINAL) — NO → CHECK LINE VOLTAGE POWER, LOW VOLTAGE TRANSFORMER, LIMIT CONTROLLER, THERMOSTAT AND WIRING. ALSO, CHECK AIR PROVING SWITCH ON COMBUSTION AIR BLOWER SYSTEM AND THAT VENT DAMPER (IF USED) IS OPEN AND END SWITCH IS MADE.

YES

30 SECOND DELAY — NO → REPLACE S860.

YES

SPARK ACROSS IGNITER/SENSOR GAP — NO → PULL IGNITION LEAD AND CHECK SPARK AT S860 IGN. STUD. SPARK OKAY ? — NO → • CHECK FUSE—REPLACE IF NECESSARY. • REPLACE S860, IF FUSE OKAY.

YES

YES →
• CHECK IGNITION CABLE, GROUND WIRING, CERAMIC INSULATOR AND GAP, AND CORRECT.
• CHECK BOOT OF THE IGNITION CABLE FOR SIGNS OF MELTING OR BUCKLING. TAKE PROTECTIVE ACTION TO SHIELD CABLE AND BOOT FROM EXCESSIVE TEMPERATURES.

TURN GAS SUPPLY ON
PILOT BURNER LIGHTS? — NO →
• CHECK THAT ALL MANUAL GAS COCKS ARE OPEN, SUPPLY TUBING AND PRESSURES ARE GOOD, AND PILOT BURNER ORIFICE IS NOT BLOCKED.
• CHECK ELECTRICAL CONNECTIONS BETWEEN S860 AND PILOT OPERATOR ON GAS CONTROL.
• CHECK FOR 24 Vac ACROSS PV—MV/PV TERMINALS ON S860; IF VOLTAGE IS OKAY, REPLACE GAS CONTROL; IF NO VOLTAGE, REPLACE S860.

YES

SPARK STOPS WHEN PILOT IS LIT? — NO →
NOTE: IF S860 GOES INTO LOCKOUT, RESET SYSTEM.
• CHECK CONTINUITY OF IGNITION CABLE AND GROUND WIRE.
• CHECK THAT PILOT FLAME COVERS ELECTRODE.
• IF CHECKS ARE OKAY, REPLACE S860 MODULE.
• ADJUST FLAME ROD SO FLAME CURRENT READING ON MICROAMMETER IS AT LEAST 1.2 μA ON S860C AND 1.0 μA ON S860D.

YES

MAIN BURNER LIGHTS? — NO →
• CHECK FOR 24 Vac (NOMINAL) ACROSS MV-MV/PV TERMINALS. IF NO VOLTAGE, REPLACE S860.
• CHECK ELECTRICAL CONNECTIONS BETWEEN S860 AND GAS CONTROL. IF OKAY, REPLACE GAS CONTROL OR GAS CONTROL OPERATOR.

YES

SYSTEM RUNS UNTIL CALL FOR HEAT ENDS? — NO →
NOTE: IF S860 GOES INTO LOCKOUT, RESET SYSTEM.
• CHECK CONTINUITY OF IGNITION CABLE AND GROUND WIRE. NOTE: IF GROUND IS POOR OR ERRATIC, SHUTDOWNS MAY OCCUR OCCASIONALLY EVEN THOUGH OPERATION IS NORMAL AT THE TIME OF CHECKOUT.
• IF CHECKS ARE OKAY, REPLACE S860 MODULE.

YES

CALL FOR HEAT ENDS
SYSTEM SHUTS OFF? — NO →
• CHECK FOR PROPER THERMOSTAT (CONTROLLER) OPERATION.
• REMOVE MV LEAD AT S860; IF VALVE CLOSES, RECHECK TEMPERATURE CONTROLLER AND WIRING; IF NOT, REPLACE GAS CONTROL.

YES

TROUBLESHOOTING ENDS   REPEAT PROCEDURE UNTIL TROUBLEFREE OPERATION IS OBTAINED.

11,192D

Figure 7.8 S860 troubleshooting guide. (Courtesy of Honeywell.)

tight. If the wire is damaged or deteriorated, replace it with 14- to 18-gauge, moisture-resistant, thermoplastic insulated wire with 105 °C (221 °F) minimum continuous rating.

C. Check the temperature at the ceramic flame rod insulator. An excessive temperature will permit leakage to ground. Provide a shield if the temperature exceeds the rating of the ignitor sensor.

D. If the flame rod or bracket is bent out of position, restore it to the correct position.

E. Replace the pilot burner/ignitor sensor with an identical unit if the insulator is cracked.

Step 3: Check the Spark Ignition Circuit. You will need a short jumper wire made from ignition cable or other heavily insulated wire.

1. Close the manual shutoff valve.
2. Disconnect the ignition cable at the stud terminl on the module.

Warning: When performing the following steps, do not touch the stripped end of the jumper wire or the stud terminal. The ignition circuit generates 15–16 kV open circuit and electrical shock can result.

---

3. Energize the module and immediately touch one end of the jumper wire firmly to the GND terminal on the module. Move the free end of the jumper slowly toward the stud terminal until a spark is established.
4. Pull the jumper wire slowly away from the stud and note the length of the gap when the sparking stops (see Table 7.2).
5. Open the manual shutoff valve and reset the system.

TABLE 7.2   ARC LENGTH (COURTESY OF HONEYWELL)

Arc length	Action
No arc or arc less than 1/8 in. [3 mm]	• Check external fuse, if provided. • Verify power at module input terminal • Replace module if fuse and power ok.
Arc 1/8 in. [3 mm] or longer	Voltage output is adequate.

Step 4: Check the Pilot Flame Current. Use the following steps to check the pilot flame current.

1. Turn off the furnace at the thermostat.
2. Disconnect the ground wire from the GND terminal on the S86.
3. Disconnect the main valve wire from the TH terminal on the gas control.

4. Set the thermostat above room temperature to call for heat. A spark will ignite the pilot, but because the main valve actuator is disconnected, the main burner will not light.

5. Make sure that the pilot flame envelopes 3/8 to 1/2 in. (10 to 13 mm) of the flame rod. If necessary, adjust the pilot flame by turning the pilot adjustment screw on the gas control clockwise to decrease or counterclockwise to increase the pilot flame. Always replace the pilot adjustment cover screw and tighten firmly to prevent gas leaks.

6. Adjust the ignitor–sensor rod on the pilot until the flame current reading on the microammeter is at the maximum, then lock it in place with the setscrew. The flame current must be at least that shown in Table 7.3.

TABLE 7.3 PILOT FLAME CURRENT READINGS
(COURTESY OF HONEYWELL)

Module	Minimum flame current
S86A, B, D, E, F, H;	
S90A, B; S860D	1.0 μA
S86C, G; S860C	1.2 μA

7. Remove the microammeter and correctly reconnect all wires. Return the system to normal operation before leaving the job.

CHECKING OUT THE DIRECT IGNITION SYSTEMS

To check out the direct ignition systems use the following steps:
Check out the gas control system:

1. At the initial installation of the appliance.

2. As a part of regular maintenance procedures. Maintenance intervals are determined by the application. See the Applications section listed earlier for more information.

3. As the first step in troubleshooting.

4. Any time work is done on the system.

> **Warning:** Fire or explosion hazard. Can cause a fire or explosion with property damage, injury, or loss of life.

1. If you smell gas or suspect a gas leak, turn off the gas at the manual service valve and evacuate the house. Do not try to light any appliance

or touch any electrical switch or telephone in the building until you are sure that no spilled gas remains.

2. A gas leak test, described in steps 1 and 5 below, must be done on an initial installation and any time work is done involving the gas piping.

Step 1: Perform a Visual Inspection.

1. With the electric power off, make sure that all wiring connections are clean and tight.
2. Turn on the electric power to the appliance and the ignition module.
3. Open the manual shutoff valves in the gas line to the appliance.
4. Do a gas leak test ahead of the gas control if the piping has been disturbed.

Gas Leak Test: Paint the pipe joints with a rich soap-and-water solution. Bubbles indicate a gas leak. Tighten the joints to stop the leak.

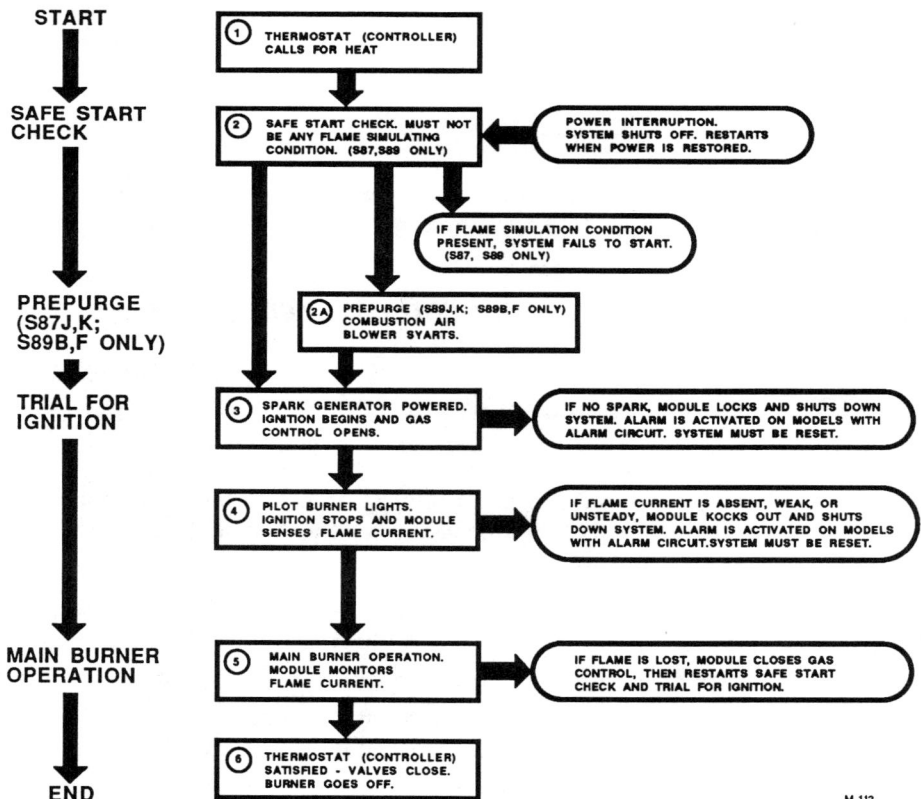

START

SAFE START CHECK

PREPURGE (S87J,K; S89B,F ONLY)

TRIAL FOR IGNITION

MAIN BURNER OPERATION

END

① THERMOSTAT (CONTROLLER) CALLS FOR HEAT

② SAFE START CHECK. MUST NOT BE ANY FLAME SIMULATING CONDITION. (S87,S89 ONLY) — POWER INTERRUPTION. SYSTEM SHUTS OFF. RESTARTS WHEN POWER IS RESTORED.

IF FLAME SIMULATION CONDITION PRESENT, SYSTEM FAILS TO START. (S87, S89 ONLY)

②A PREPURGE (S89J,K; S89B,F ONLY) COMBUSTION AIR BLOWER SYARTS.

③ SPARK GENERATOR POWERED. IGNITION BEGINS AND GAS CONTROL OPENS. — IF NO SPARK, MODULE LOCKS AND SHUTS DOWN SYSTEM. ALARM IS ACTIVATED ON MODELS WITH ALARM CIRCUIT. SYSTEM MUST BE RESET.

④ PILOT BURNER LIGHTS. IGNITION STOPS AND MODULE SENSES FLAME CURRENT. — IF FLAME CURRENT IS ABSENT, WEAK, OR UNSTEADY, MODULE KOCKS OUT AND SHUTS DOWN SYSTEM. ALARM IS ACTIVATED ON MODELS WITH ALARM CIRCUIT.SYSTEM MUST BE RESET.

⑤ MAIN BURNER OPERATION. MODULE MONITORS FLAME CURRENT. — IF FLAME IS LOST, MODULE CLOSES GAS CONTROL, THEN RESTARTS SAFE START CHECK AND TRIAL FOR IGNITION.

⑥ THERMOSTAT (CONTROLLER) SATISFIED - VALVES CLOSE. BURNER GOES OFF.

M 112

Figure 7.9 S825, S87, S89A, B, E, F normal operating sequence. (Courtesy of Honeywell.)

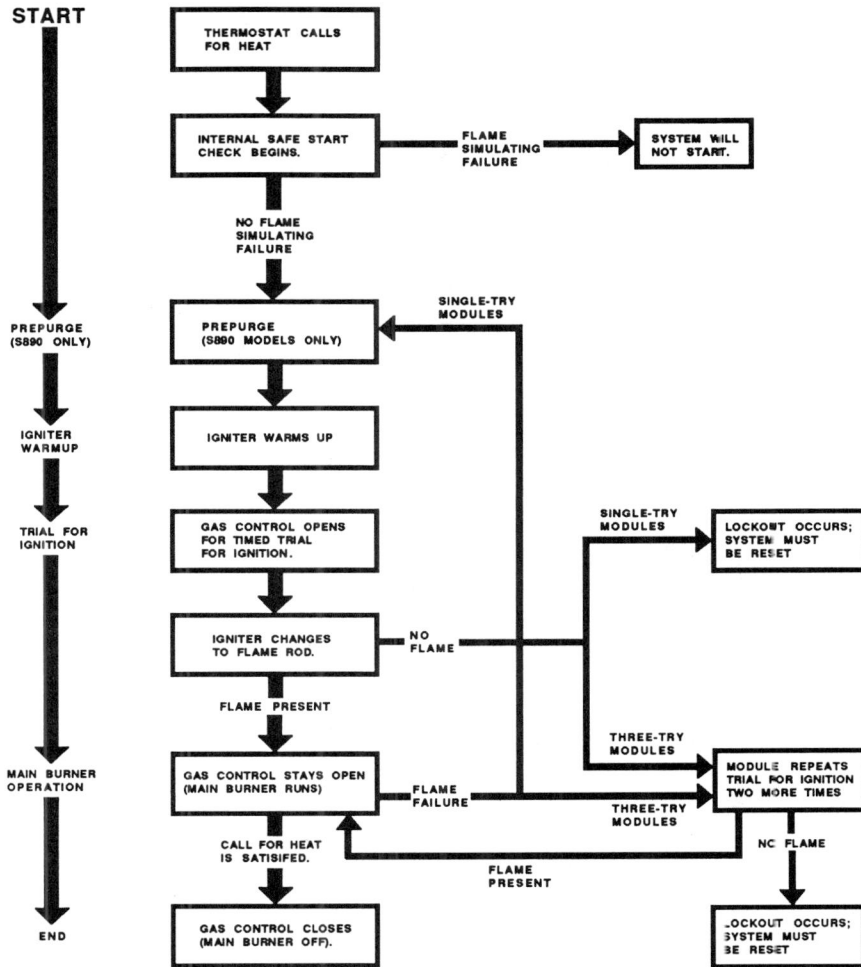

Figure 7.10 S89 C, D, G, H normal operating sequence. (Courtesy of Honeywell.)

Step 2: Review the normal operating sequence and timing summary. (See Figures 7.9 and 7.10 and Table 7.4).

Step 3: Reset the Module.

1. Turn the thermostat to its lowest setting.
2. Wait one minute.

As you do steps 4 and 5, watch for points where the operation deviates from normal. Refer to the Troubleshooting Chart to correct the problem.

TABLE 7.4 DIRECT IGNITION MODULE TIMING SUMMARY (COURTESY OF HONEYWELL)

Function	S825	S87A-D	S87J,K	S89A,E	S89B,F	S89C,D	S89G,H	S890C,D	S890G,H
Prepurge	—	—	30 sec. min.	—	30 sec. min.; 45 sec. max.	—	—	30 sec. min.; 37 sec. max.	30 sec. min.; 37 sec. max.
No. of ignition trials	1	1	1	1	1	1	3	1	3
Safety lockout timing (nom.)[a]	6, 11 or 21 sec.	4, 6, 11 or 21 sec	4, 6, 11 or 21 sec.	4, 6, 11, 15[b] or 21 sec	4, 6, 11, 15[b] or 21 sec.	4, 6, 11, 15 or 21[c] sec.	4, 6, 11 or 15 sec.	4, 6, 11 or 15 sec.	4, 6, 11 or 15 sec.
Flame failure response	0.8 sec. max.	0.8 sec. max	0.8 sec. max.	0.8 sec. max.	0.8 sec. max	2 sec. max.[d]	2 sec max.[d]	2 sec. max.[d]	2 sec. max.[d]
Igniter warmup time	—	—	—	—	—	34 sec.[e]	34 sec.	34 sec.	34 sec.

[a]Ignition continues until burner lights or system locks out.
[b]S89E,F only.
[c]S89C before 9/87 only.
[d]With 2.5 μA flame signal.
[e]S89C before 9/87 warmup period was 45 sec. with 2.5 μA flame signal.

Step 4: Check the Safety Lockout Operation.
1. Turn the gas supply off.
2. Set the thermostat or the controller above room temperature to call for heat.
3. Watch for sparking at the ignitor immediately or following a prepurge (see the timing summary, Table 7.4).
4. Check the lockout timing.
5. After the lockout, open the manual gas cock and make sure that no gas is flowing to the main burner (or through the enrichment tube if the Q366 is used).
6. Set the thermostat below room temperature and wait one minute before continuing.

Step 5: Check the Normal Operation.
1. Set the thermostat or controller above room temperature to call for heat.
2. Make sure that the main burner lights smoothly without flashback. Several attempts may be necessary to clear the gas line of air.
3. Make sure that the burner operates smoothly without floating, lifting, or flame rollout to the furnace vestibule or heat buildup in the vestibule.
4. If the gas line has been disturbed, complete a gas leak test.

> **Gas Leak Test:** Paint the gas control gasket edges and all pipe connections downstream of the gas control, including the enrichment tube connections, with a rich soap-and-water solution. Bubbles indicate gas leaks. Tighten the joints and screws or replace the component to stop the gas leak.

5. Turn the thermostat or the controller below room temperature. Make sure that the main burner flame goes out.

TROUBLESHOOTING DIRECT IGNITION SYSTEMS

Follow the appropriate troubleshooting guide to pinpoint the cause of the problem (see Figures 7.11 and 7.12). If troubleshooting indicates an ignition problem, see the Ignition System Checks below to isolate and correct the problem.

Following the troubleshooting, perform the checkout procedure outlined above again to be sure that the system is operating normally.

Ignition system checks. Use the following procedures when making ignition system checks:

START

NOTE: BEFORE TROUBLESHOOTING, FAMILIARIZE YOURSELF WITH THE STARTUP AND CHECKOUT PROCEDURE.

REVIEW NORMAL OPERATING SEQUENCE. TURN THERMOSTAT UP TO CALL FOR HEAT. TURN GAS SUPPLY ON.

POWER TO MODULE S87-24V NOM. S89-24V AND 120V NOM.

NO → CHECK LINE VOLTAGE POWER, LOW VOLTAGE TRANSFORMER, LIMIT CONTROLLER, THERMOSTAT AND WIRING. ALSO, AIR PROVING SWITCH ON PREPURGE SYSTEMS (SEE WIRING HOOKUPS).

YES

S89E - 8 SEC. DELAY FOR SAFE START CHECK S87J,K; S89F 30 SECOND DELAY FOR PREPURGE

NO → REPLACE S89.

YES

SPARK ACROSS IGNITER OR IGNITER/SENSOR GAP

NO → CHECK SPARK IGNITION CIRCUIT. SEE COMPONENT CHECKS.

SPARK OKAY?

NO →
- CHECK FOR 120V TO SPARK GENERATOR (S89).
- CHECK EXTERNAL FUSE (S87) AND REPLACE IF NECESSARY.
- REPLACE SPARK GENERATOR IF VOLTAGE OK.
- IF NO VOLTAGE, REPLACE MODULE.

YES

- MAKE SURE IGNITION CABLE
  — PROVIDES ELECTRICAL CONTINUITY.
  — DOES NOT TOUCH ANY METAL SURFACES.
  — CONNECTIONS ARE CLEAN AND TIGHT.
  — SHOWS NO SIGNS OF MELTING OR BUCKING.
  REPLACE AND SHIELD CABLE IF NECESSARY.
- MAKE SURE BURNER, SPARK IGNITER AND MODULE GND (BURNER) TERMINAL HAVE EFFECTIVE COMMON GROUND. POOR OR ERRATIC GROUND WILL CAUSE NUISANCE SHUTDOWNS.
- CHECK FOR CRACKED INSULATOR ON IGNITER OR FLAME SENSOR. REPLACE DEVICE WITH CRACKED INSULATOR TO PREVENT SHORT TO GROUND.

YES

MAIN BURNER LIGHTS

NO → NOTE: IF MODULE LOCKS OUT, RESET BEFORE CONTINUING.[a]
- CHECK FOR 24 Vac ACROSS VALVE AND VALVE (GND) TERMINALS ON MODULE (ON S825A,C. CHECK FOR 9±1 VDC ACROSS 5 AND 3). IF NO VOLTAGE, REPLACE MODULE.
- MAKE SURE IGNITER AND SENSOR ARE PROPERLY POSITIONED. SEE COMPONENT CHECKS.
- CHECK ELECTRICAL CONNECTIONS BETWEEN MODULE AND GAS CONTROL. IF OKAY, REPLACE GAS CONTROL.

YES

SPARK STOPS WHEN BURNER IS LIT

NO → SPARK CONTINUES AFTER BURNER IS LIT

NO → NOTE: IF MODULE GOES INTO LOCKOUT, RESET SYSTEM.[a]
- CHECK CONTINUITY OF SENSOR CABLE AND GROUND WIRE.
- CHECK THAT BURNER FLAME COVERS ALL ELECTRODES.
- IF CHECKS ARE OKAY, REPLACE MODULE.
- ON S87, REVERSE LEADS TO GAS CONTROL. IF SPARK CONTINUES, REPLACE WITH S87 BUILT AFTER 3/87.

YES

SPARK STOPS BEFORE END OF IGNITION TIMING (NORMAL ON S89E,F).

NO → REPLACE MODULE.

SYSTEM RUNS UNTIL CALL FOR HEAT ENDS

NO → NOTE: IF MODULE LOCKS OUT, RESET BEFORE CONTINUING.[a]
- MAKE SURE FLAME CURRENT IS CORRECT. SEE COMPONENT CHECKS.
- MAKE SURE L1 AND L2 ON S89 ARE CONNECTED TO THE PROPER TERMINALS.
- CHECK FOR EXCESSIVE HEAT AT SENSOR INSULATOR (TEMPERATURE ABOVE 1000 F [538 C] CAUSES SHORT TO GROUND).
- IF CHECKS ARE OKAY, REPLACE MODULE.

YES

CALL FOR HEAT ENDS; SYSTEM SHUTS OFF

NO →
- CHECK FOR PROPER TEMPERATURE CONTROLLER OPERATION.
- REMOVE VALVE LEAD AT MODULE; IF VALVE CLOSES, RECHECK TEMPERATURE CONTROLLER AND WIRING; IF NOT, REPLACE GAS CONTROL.

YES

[a] ON MODELS WITH ALARM, ALARM CIRCUIT MAKES TO TURN ON ALARM WHEN LOCKOUT OCCURS.

TROUBLESHOOTING ENDS

REPEAT PROCEDURE UNTIL TROUBLEFREE OPERATION IS OBTAINED.

M 109

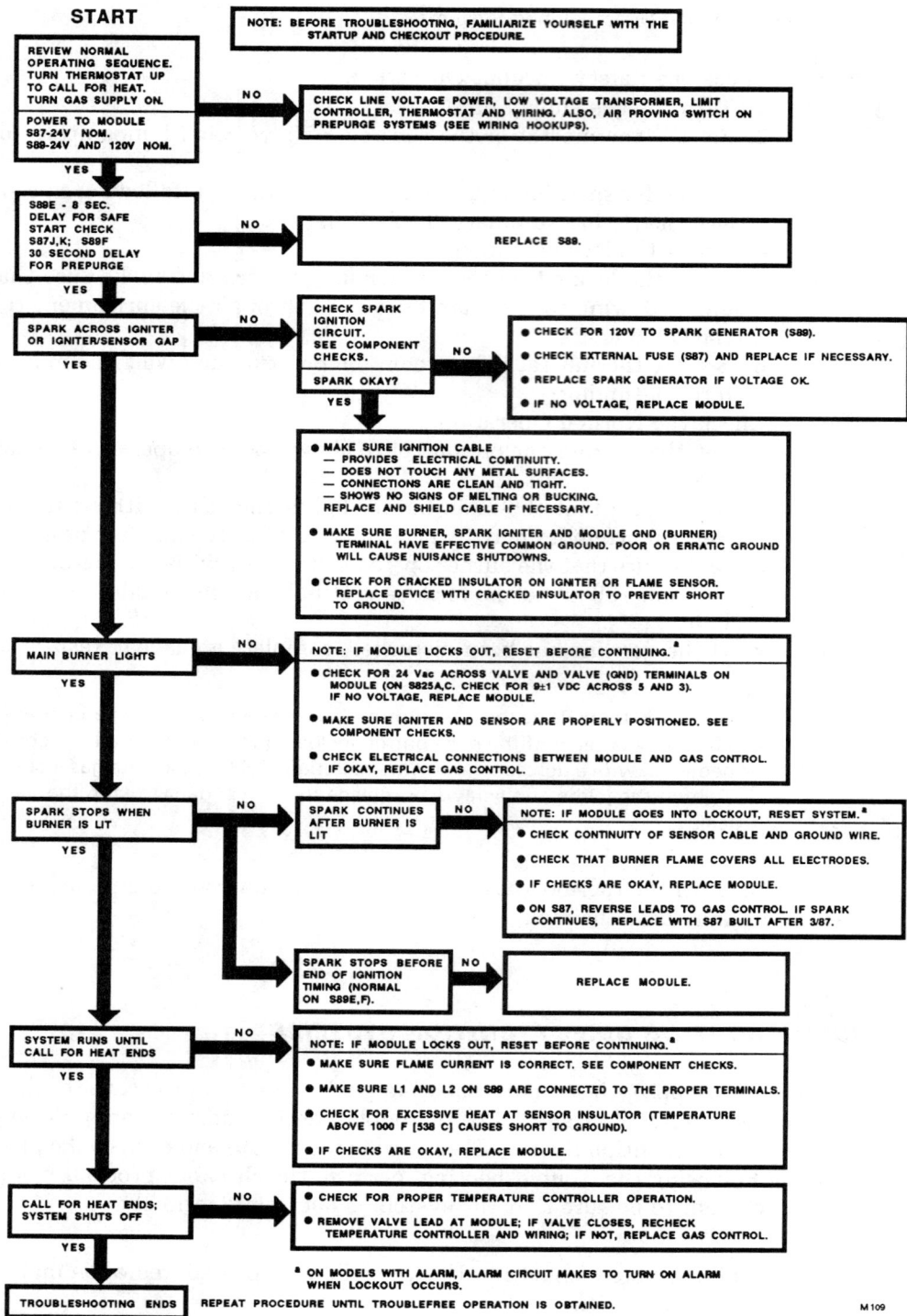

Figure 7.11 S825; S87A, B, E, F troubleshooting guide. (Courtesy of Honeywell.)

START

TURN THERMOSTAT TO CALL FOR HEAT

NOTE: BEFORE TROUBLESHOOTING, FAMILIARIZE YOURSELF WITH START-UP AND CHECKOUT PROCEDURE.

DOES S89/S890 GET POWER (24 VAC NOMINAL)? — NO →
- CHECK LINE VOLTAGE POWER.
- CHECK LOW VOLTAGE TRANSFORMER.
- CHECK LIMIT CONTROLLER.
- CHECK AIR PROVING SWITCH (IF USED).
- CHECK THERMOSTAT.
- CHECK WIRING.

YES ↓

34 SEC PREPURGE S890 MODELS ONLY — NO →
- CHECK THAT CONTROL IS S890 TYPE.
- CHECK AIR PROVING SWITCH.
- REPLACE S890.

YES ↓

IGNITER WARMS UP AND GLOWS RED — NO →
CHECK FOR 120 Vac TO IGNITER. — NO → REPLACE S89/S890.

YES ↓ REPLACE IGNITER.

YES ↓

MAIN BURNER LIGHTS — NO →
- CHECK FOR 24 Vac ACROSS VALVE AND VALVE TERMINALS ON S89/S890 DURING LOCKOUT TIME. IF NO VOLTAGE, REPLACE S89/S890.
- CHECK IGNITER POSITION.
- CHECK ELECTRICAL CONNECTIONS BETWEEN S89/S890 AND GAS CONTROL. IF OKAY, REPLACE GAS CONTROL.

YES ↓

MAIN BURNER REMAINS POWERED AND LIT — NO →
NOTE: IF S89/S890 GOES INTO LOCKOUT, RESET SYSTEM.
- CHECK CONTINUITY OF GROUND WIRE.
- MAKE SURE L1 AND L2 ARE NOT REVERSED; THIS WOULD PREVENT FLAME DETECTION.
- CHECK THAT BURNER FLAME COVERS IGNITER.
- CHECK INSULATION ON IGNITER LEADS (C AND G MODELS).
- CHECK INSULATION ON SENSOR LEADS AND CHECK SENSOR POSITION (D AND H MODELS).
- IF CHECKS ARE OKAY, REPLACE S89/S890 MODULE.

YES ↓

SYSTEM RUNS UNTIL CALL FOR HEAT ENDS — NO →
NOTE: IF S89/S890 GOES INTO LOCKOUT, RESET SYSTEM.
- CHECK CONTINUITY OF GROUND WIRE.
  NOTE: IF GROUND IS POOR OR ERRATIC, SHUTDOWNS MAY OCCUR OCCASIONALLY EVEN THOUGH OPERATION IS NORMAL AT THE TIME OF CHECKOUT.
- CHECK FOR EXCESSIVE HEAT AT IGNITER CERAMIC BASE (TEMPERATURE ABOVE 1000° F [538° C] CAUSES SHORT TO GROUND).
- IF CHECKS ARE OKAY, REPLACE S89/S890 MODULE.

YES ↓

CALL FOR HEAT ENDS: SYSTEM SHUTS OFF — NO →
- CHECK FOR PROPER TEMPERATURE CONTROLLER OPERATION.
- REMOVE VALVE LEAD AT S89/S890; IF VALVE CLOSES, RECHECK TEMPERATURE CONTROLLER AND WIRING; IF NOT, REPLACE GAS CONTROL.

YES ↓

TROUBLESHOOTING ENDS

REPEAT PROCEDURE UNTIL TROUBLEFREE OPERATION IS OBTAINED.

12,334A

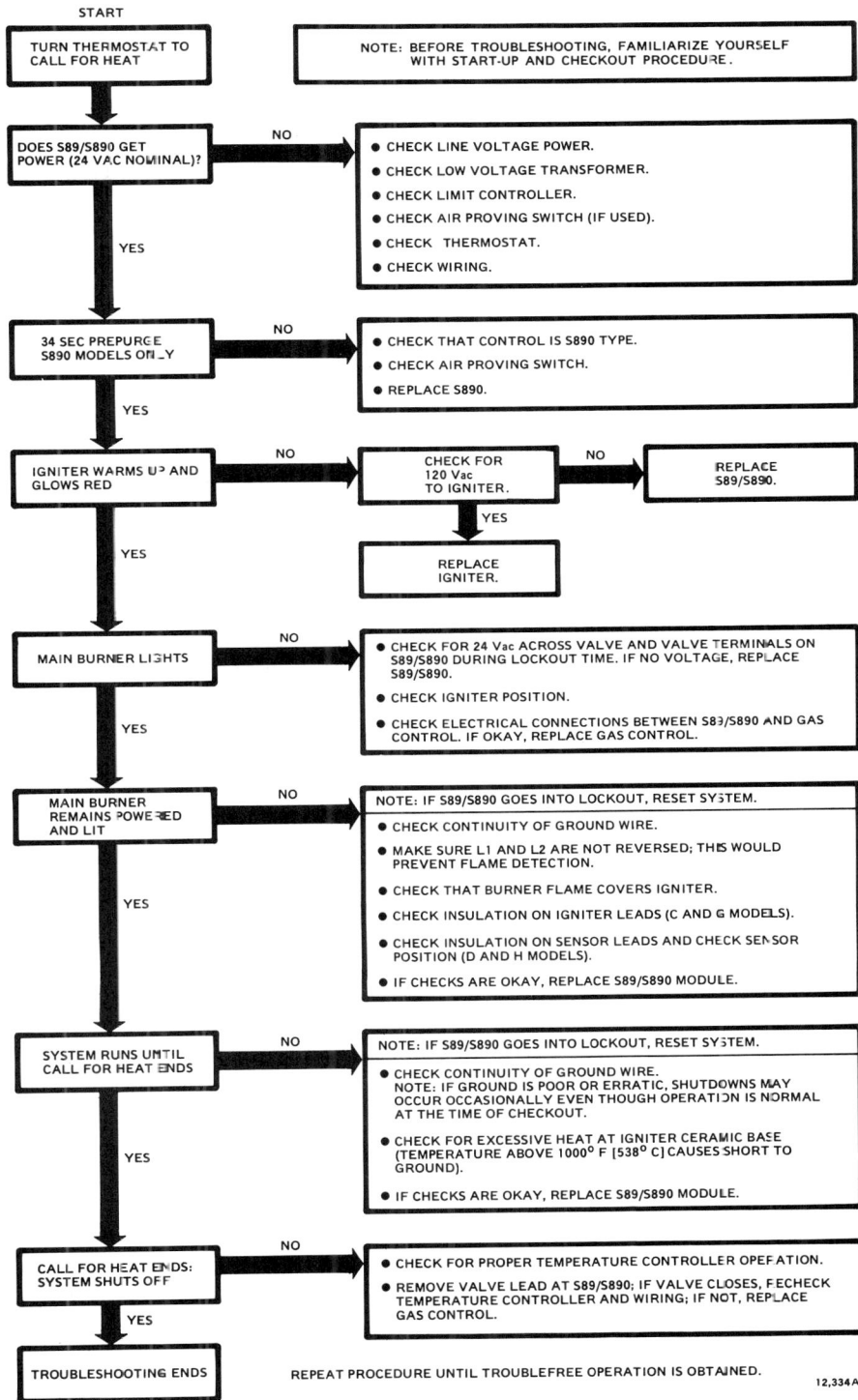

Figure 7.12 S89C, D, G, H troubleshooting guide. (Courtesy of Honeywell.)

Step 1: Check the Ignition Cable.

Make sure that the:

1. Ignition cable does not touch any metal surface.
2. Ignition cable is no more than 36 in. (0.9 m) long.
3. Connections to the stud terminal and ignitor–sensor are clean and tight.
4. Ignition cable provides good electrical continuity.

Step 2: Check the Ignition System Grounding. Nuisance shutdowns are often caused by a poor or erratic ground.

1. A common ground is required for the module, ignitor, flame sensor, and main burner.

   A. Check for good metal-to-metal contact between the ignitor bracket and the main burner.

   B. Check the ground lead from the GND (burner) terminal on the module to the ignitor bracket. Make sure that the connections are clean and tight. If the wire is damaged or deteriorated, replace it with 14- to 16-gauge, moisture-resistant, thermoplastic insulated wire with 105 °C (221 °F) minimum rating. Use a shield if necessary to protect the ground wire from radiant heat. Check the temperature at the ceramic ignitor–sensor insulator. Excessive temperature will permit leakage to ground. Provide a shield if the temperature exceeds the rating of the ignitor or sensor.

   C. If the flame sensor or bracket is bent out of position, restore it to the correct position.

   D. Replace the ignitor and sensor or ignitor–sensor with an identical unit if the insulator is cracked.

Step 3: Check the Spark Ignition Circuit. You will need a short jumper wire made from ignition cable or other heavily insulated wire.

1. Close the manual shutoff valve.
2. Disconnect the ignition cable at the stud terminal on the module or the spark generator.

> **Warning:** When performing the following steps, do not touch the stripped end of the jumper or the stud terminal. The ignition circuit generates 24 kV open circuit and an electrical shock can result.

3. Energize the module and immediately touch one end of the jumper firmly to the GND terminal on the module. Move the free end of the jumper slowly toward the stud terminal until a spark is established.
4. Pull the jumper slowly away from the stud and note the length of the gap when the sparking stops (see Table 7.5).
5. Open the manual shutoff valve and reset the system.

TABLE 7.5   ARC LENGTH (COURTESY OF HONEYWELL)

Arc length	Action
No arc or arc less than 1/8 in. [3 mm] (3/16 in. [5 mm] on S825)	• Check external fuse, if provided. • Verify power at module input terminals, or at spark generator input terminals • Replace module and/or spark generator if fuse and power ok.
Arc 1/8 in. [3 mm] (3/16 in. [5 mm] on S825) or longer	Voltage output is adequate.

Step 4: Check the Flame Sensor Circuit. These steps are for single-rod S87 DSI only.

  1. Turn off the furnace at the thermostat.
  2. Connect a meter (dc microammeter scale) in series with the ground lead as shown in Figure 7.13. Use the Honeywell W136 test meter or equivalent. Connect the meter as follows:
     A. Disconnect the ground lead at the electronic control.
     B. Connect the black (negative) meter lead to the electronic control GND terminal.
     C. Connect the red (positive) meter lead to the free end of the ground lead.
  3. Restart the system and read the meter. The flame sensor current must be steady and at least 1.5 μA.
  4. If the meter reads less than the minimum or the reading is unsteady, complete the following steps:
     A. Make sure that the burner flame is capable of providing a good rectification signal (see Figure 7.14).
     B. Make sure that about 1 in. of the flame sensor or ignitor–sensor is continuously immersed in the flame for the best flame signal. Turn off the system and allow it to cool, then bend the bracket or flame sensor, or relocate the sensor as necessary. Do not relocate an ignitor–sensor.
     C. Check the excessive (over 1000 °F or 538 °C) temperature at the ceramic insulator on the flame sensor. Excessive temperature can cause a short to ground; move the sensor to a cooler location or shield the insulator. Do not relocate an ignitor–sensor.
     D. Check for a cracked ceramic insulator, which can cause a short to ground, and replace the sensor if necessary.
     E. Make sure that the electrical connections are clean and tight. Replace damaged wire with moisture-resistant No. 18 wire rated for continuous duty up to 105 °C (221 °F).
  5. Remove the microammeter and reconnect the ground wire. Return the system to normal operation.

Checking Two-Rod S87 and S89.

FLAME SENSOR CURRENT CHECK—USE µA SCALE

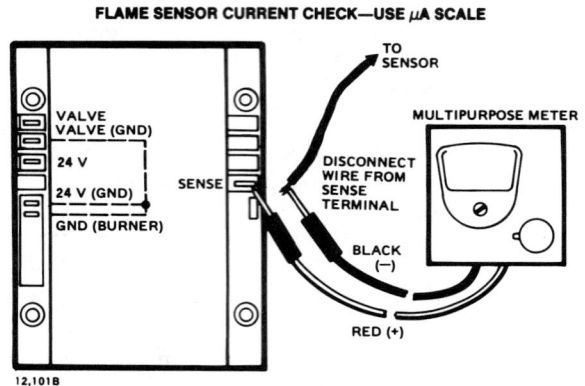

12,101B

Checking Single Rod S87.

FLAME SENSOR CURRENT CHECK—USE µA SCALE

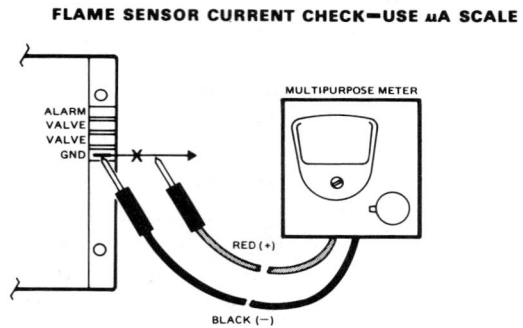

Figure 7.13 Checking flame current on S87 and S89. (Courtesy of Honeywell.)

These steps are for two-rod, DSI systems only:

1. Turn off the furnace at the thermostat.
2. Connect a meter (dc microammeter) in series with the flame sensor lead (see Figure 7.13 or 7.15). Use the Honeywell W136 test meter or equivalent. Connect the meter as follows:
   A. Disconnect the sensor lead at the electronic control.
   B. Connect the red (positive) meter lead to the electronic SENSE terminal.
   C. Connect the black (negative) meter lead to the free end of the sensor lead.
3. Restart the system and read the meter. The flame sensor current must be steady and at least the minimum as shown in Table 7.6.

CHECK BURNER FLAME CONDITIONS.

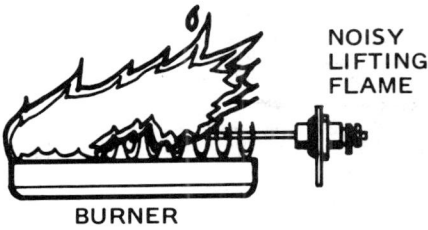

NOISY
LIFTING
FLAME

CHECK FOR:
- HIGH GAS PRESSURE
- EXCESS PRIMARY AIR OR DRAFT

BURNER

WAVING
FLAME

CHECK FOR:
- POOR DRAFT
- EXCESS DRAFT
- HIGH VELOCITY OF SECCNDARY
  AIR
INSTALL SHIELD IF NECESSARY.

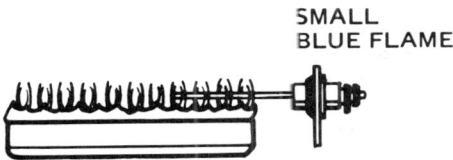

SMALL
BLUE FLAME

CHECK FOR:
- CLOGGED PORTS OR ORIFICE
  FILTER
- WRONG SIZE ORIFICE

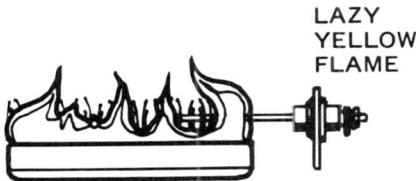

LAZY
YELLOW
FLAME

CHECK FOR:
LACK OF AIR FROM
- DIRTY PRIMARY AIR
  OPENING
- LARGE PORTS OR
  ORIFICES

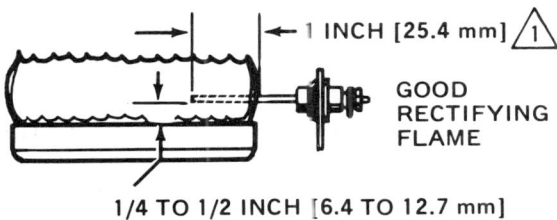

← 1 INCH [25.4 mm] ⚠1

GOOD
RECTIFYING
FLAME

1/4 TO 1/2 INCH [6.4 TO 12.7 mm]

⚠1 3/4 TO 1 INCH ON HSI IGNITER   12,616

Figure 7.14 Burner flame must provide good current path. (Courtesy of Honeywell.)

S825C SYSTEM CONTROL

S825D OUTPUT RELAY CIRCUIT
(OTHER METER CHECKS SAME AS FOR S825C)

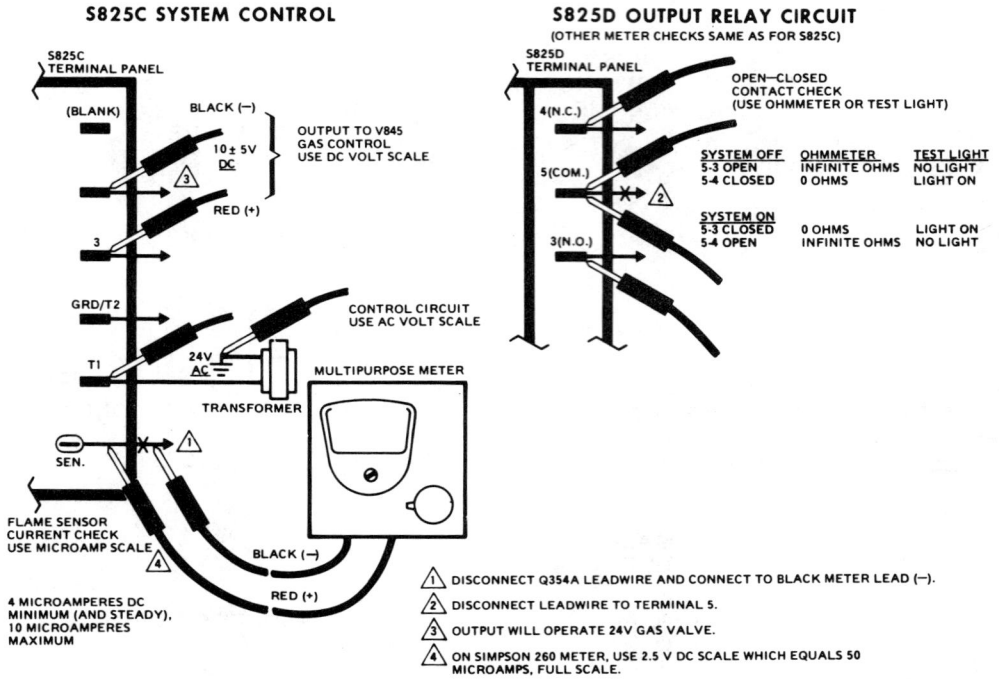

Figure 7.15 Checking S825. (Courtesy of Honeywell.)

TABLE 7.6 PILOT FLAME CURRENT READINGS
(COURTESY OF HONEYWELL)

Module	Minimum flame current
S87, S89A, B	1.5 μA
S89E, F	0.8 μA
S825	4.0 μA (Max. current is 10.0 μA).

4. If the meter reads less than the minimum or the reading is unsteady use the following procedure:
   A. Make sure that the burner flame is capable of providing a good rectification signal (see Figure 7.14).
   B. Make sure that about 1 in. of the flame sensor or ignitor to sensor is continuously immersed in the flame for the best signal (see Figure 7.14). Turn off the system and allow it to cool, then bend the bracket or flame sensor, or relocate the sensor as necessary. Do not relocate an ignitor or combination ignitor–sensor.
   C. Check for excessive (over 1000 °F or 538 °C) temperature at the ceramic insulator on the flame sensor. Excessive temperature can cause

a short to ground; move the sensor to a cooler location or shield the insulator. Do not relocate an ignitor–sensor.

    D. Check for a cracked ceramic insulator, which can cause a short to ground, and replace the sensor if necessary.

    E. Make sure that the electrical connections are clean and tight. Replace damaged wire with moisture-resistant No. 18 wire rated for continuous duty up to 105 °C (221 °F).

5. On the S85, if the reading is over 10 μA, move the Q354 flame sensor so that less of the flame rod is immersed in the flame.

6. Remove the microammeter and reconnect the sensor lead. Return the system to normal operation.

These steps are for HSI (hot surface ignition) systems only.

1. Make sure that the burner flame is capable of providing a good rectification signal (see Figure 7.14).

2. Make sure that about 3/4 to 1 in. of the flame sensor or ignitor–sensor is continuously immersed in the flame for the best flame signal. Bend the bracket or the flame sensor, or relocate the sensor as necessary. Do not relocate an ignitor or combination ignitor–sensor.

3. Check for excessive (over 1000 °F or 538 °C) temperature at the ceramic insulator on the flame sensor. Excessive temperature can cause a short to ground; move the sensor to a cooler location or shield the insulator. Do not relocate an ignitor or combination ignitor–sensor.

4. Check for a cracked ceramic insulator, which can cause a short to ground, and replace the sensor if necessary.

    A. Make sure that the electrical connections are clean and tight. Replace damaged wire with moisture-resistant No. 18 wire rated for continuous duty up to 105 °C (221 °F).

5. If the ignitor is other than a Norton 201 or Honeywell 201071, make sure that it meets the following specifications:

    A. The ignitor must reach 1000 °C (1832 °F) within 34 seconds with 102 VAC applied.

    B. The ignitor must maintain at least 500 000 Ω insulation resistance between the ignitor leadwires and the ignitor mounting bracket.

    C. The ignitor must develop an insulating layer on its surface (over time) that would prevent flame sensing.

    D. The ignitor surface area immersed in the flame must not exceed 1/4 of the grounded area immersed in the flame. This would prevent flame sensing.

    E. The ignitor current draw at 132 VAC must not exceed 5 A.

For troubleshooting procedures for the Honeywell spark to pilot (IID) system see the following chart.

HONEYWELL IID TROUBLESHOOTING CHART

HONEYWELL SPARK TO PILOT (I.I.D.) TROUBLE SHOOTING CHART

Symptoms	Possible cause	Checks & remedies
No spark	Open in ignition cable.	Check continuity of ignition cable if open, replace.
	Ignitor improperly grounded.	Check ground connections, insure good chasis ground.
	No voltage to ignition module.	Verify 25 volts AC input to ignition module.
	Ignitor improperly adjusted.	Verify correct ignitor adjustment.
	Defective control module.	Check for spark from module.
Spark, but no ignition	No gas to pilot assembly.	Verify supply pressure.
	Pilot orifice plugged.	Inspect orifice clean or replace if dirty.
	Gas supply tubing to pilot kinked.	Inspect tubing to pilot assembly correct or replace.
	No voltage to gas control.	Verify voltage across terminals PV-MV/PV of module. If no voltage, replace module, if present replace gas control.
Spark continues after ignition	Ignition cable has no continuity.	Check ignition cable for continuity.
	Poor flame impingement on sensor rod.	Check that flame covers both electrodes.
Main burner fails to light	No voltage to gas valve.	Verify voltage at terminals MV-MV/PV of modules. If no voltage, replace module.
	Defective gas valve.	Check connections between module and gas valve, if O.K., replace gas valve.
Main burner shuts down	Unit improperly grounded.	Verify ignitor assembly and module is grounded properly.
Main burner fails to shut down at end of cycle	Defective thermostat.	Check thermostat.
	Defective ignition module.	Remove MV lead from module, if gas valve closes, replace module, if not replace gas valve.

Glossary

Portions of this section are presented with the permission of Heil Heating and Cooling Products, Lavergne, TN.

Absolute humidity. The weight of water vapor in a given amount of air in grains per cubic foot.

Absolute temperature. A temperature scale expressed in degrees Fahrenheit (F) or Celsius (C) [also called centigrade] using absolute zero as a base. Referred to as the Rankin or Kelvin scale.

Absolute zero. A temperature at which molecular activity theoretically ceases. With temperatures of -459.69 °F or -273.16 °C.

Air: Atmospheric air is composed of approximately 78% nitrogen, 21% oxygen, and 1% inert gases including carbon dioxide, krypton, neon, ozone, and ammonia. Over the sea, traces of salt are present, and over land, traces of sulfates are present. Dust and microorganisms are also present.

Air conditioning. The process of controlling the temperature, humidity, cleanliness, and distribution of the air.

Atmospheric gas burner. A burner that is entirely dependent upon atmospheric pressure for the necessary combustion air.

Aldehyde: A product of the primary combustion process. Chemically related to the alcohol family.

Atmospheric pressure.   The weight of one square inch of a column of the earth's atmosphere. At sea level the pressure is 14.696 pounds per square inch.

Bimetal.   Two metals with different rates of expansion fastened together. When heated or cooled they will warp and can be made to open or close a switch or valve.

British Thermal Unit (Btu).   The amount of heat necessary to change the temperature of one pound of pure water 1 °F.

Carbon dioxide.   An odorless, colorless gas formed by the complete combustion of carbon.

Carbon monoxide.   A product of incomplete combustion. Odorless and colorless, but extremely toxic. It has greater affinity for the hemoglobin of the blood than oxygen and will thereby replace it. A concentration as low as 1 point per 1000 is dangerous to life in 1/2 to 1 hour; 1 part to 10 000 is the maximum permissible concentration.

Centigrade (Celsius).   A temperature scale with the freezing point of water at 0° and a boiling point of 100° at atmospheric pressure.

Combination fan and limit switch.   A combination of the two switches operated by a common bimetal. See Fan switch and Limit switch.

Combustion.   The oxidation of a substance at a rapid rate to produce heat and some light.

Combustible fuel.   A substance that combines readily with oxygen.

Conduction.   The transfer of heat from molecule to molecule within a substance.

Convection.   The transfer of heat by a moving fluid.

Cycle.   The complete course of operation of a refrigerant back to a selected starting point in a system. Also used to describe alternating current through 360 space degrees.

Draft diverter.   A device used to regulate the draft in a gas appliance. It is a nonadjustable device specifically designed for a particular appliance. It also serves to prevent down draft into the combustion chamber area.

Draft gauge.   A manometer calibrated to read very small pressures in inches of water.

Dry bulb temperature.   The temperature read with an ordinary thermometer.

Excess air.   Air beyond that needed for theoretically perfect combustion.

Fahrenheit.   A temperature scale with the freezing point of water at 32 °F and the boiling point at 212 °F at atmospheric pressure.

Fan switch.   A bimetallic switch used to start and stop the blower on a forced air furnace. A normally open switch that closes on a temperature rise. Fan ON and OFF settings may both be adjustable or one of the two adjustable with a fixed or adjustable differential.

Gas pressure.   The pressure measured at the outlet of the valve or at a tapping in the manifold. For natural gas it is 3.5 in. wc and for LP gases 10 or 11 in. wc.

Gas valve.   An electrically actuated valve used to control the flow of gas to the main burners. Most have integral gas cocks, a pilot safety valve, and pressure regulators.

Heat.   A form of energy causing the agitation of molecules within a substance.

Heat anticipator.   A resistance heater (usually variable) in electrical series with the heating circuit in the thermostat. It adds a small amount of heat during the heating cycle and causes the thermostat to open the circuit slightly ahead of the set point. This compensates for residual heat in the heat exchanger or time delay on electric heat and prevents thermostat overshoot.

Heat exchanger.   A device for the transfer of heat energy from the source to the conveying medium.

Heat flow.   Heat flows from a warmer to a cooler substance. The rate depends upon the temperature difference, the area exposed, and the type of material.

Heat transfer.   The three methods of heat transfer are conduction, convection, and radiation.

Hydrocarbon fuel.   A substance mainly composed of hydrogen and carbon.

Inches of mercury.   Atmospheric pressure is equal to 29.92 in. of mercury.

Intermittent ignition.   Ignition that only occurs at the start of the heating cycle.

Latent heat.   Heat that produces a change in temperature, that is, ice to water at 32 °F, water to steam at 212 °F.

Limit switch.   A bimetallic switch used to limit the maximum temperature of a heating device by interrupting the control circuit to the gas valve. It is a normally closed switch and opens on a rise in temperature. Most have a fixed differential of 25 °F.

Manometer.   A tube filled with a liquid used to measure pressures.

Orifice.   A device used with gas burners to meter the flow of fuel and induce the flow of primary air.

Pilot safety valve.   A valve that will close when a pilot flame failure occurs, shutting off the gas supply to the main burner and pilot burner. Usually integral to the gas valve.

Pressure drop.   The decrease in pressure due to friction, of a fluid or vapor as it passes through a tube or duct.

Pressure regulator.   A device used to regulate the manifold gas pressure to a gas burner. Atmospheric pressure and an adjustable spring pressure on one side of a diaphragm opposes valve opening. Gas pressure on the other side will open the valve port. An additional spring pressure will result in an increase in the manifold pressure. Some are part of a gas valve.

Primary air.   Air mixed with the fuel prior to initial combustion.

Primary combustion. The ignition and initial oxidation of combustible fuel producing carbon monoxide, aldehydes, and the release of hydrogen.

Psychrometer. A device having both a dry bulb and a wet bulb thermometer. It is used to determine the relative humidity in a conditioned space. Most have an indexed scale to allow direct conversion from the temperature readings to the percentage of relative humidity.

Psychrometric chart. A chart on which can be found the properties of air under varying conditions of temperature, water vapor content, volume, etc.

Radiation. The transfer of heat without an intervening medium. It is absorbed on contact with a solid surface.

Related humidity. The percentage of water vapor present in a given quantity of air compared to the amount it can hold at this temperature.

Relay. A device used to open and/or close an electrical circuit. The relay may be actuated by a bimetal electrically heated strip, a rod wrapped with a fine resistance wire, causing expansion when energized, a bellows actuated by the expansion of a fluid or gas, or an electromagnetic coil.

Secondary air. Air mixed with the fuel during combustion and needed to complete the support of combustion.

Secondary combustion. The final combustion process converting carbon monoxide, aldehydes, and hydrogen to carbon dioxide and water vapor.

Sensible heat. Heat that can be measured or felt. Sensible heat always causes a temperature rise.

Servo pressure regulator. A small diaphragm with adjustable spring loading senses the outlet pressure opening or restricting a port and causing the main valve diaphragm to throttle regulating the outlet pressure. Servo regulators give much better regulation at varying inlet pressures and rates of flow.

Specific gravity. The weight of mass as compared to air or water which has the value of 1.

State condition. Substances can exist in three states—solid, liquid, and vapor.

Step opening valves. Gas valves designed to open partially for a short period to allow smoother ignition before opening to full input.

Temperature. A measure of the intensity of heat.

Thermocouple. Two dissimilar metals joined together at each end. When one junction is maintained at a higher temperature a low-voltage direct current will flow. Used to operate pilot safety valves. Thermocouples have an output in the range of 30 mV.

Thermostat. A bimetal-actuated switch used to close or open a circuit to initiate or terminate operation of a heating or cooling system.

Total heat (enthalpy). The total heat energy in a substance. The sum of the sensible and latent heat.

Two-stage pressure regulation. Used with LP gases. The main tank regulator is set to a value higher than 11 in. wc and each appliance has an individual regulator to maintain 11 in. wc pressure.

Vacuum. Any pressure below atmospheric pressure.

Wet bulb temperature. The temperature read with a thermometer whose bulb is encased in a wetted wick.

Index